Practical Handbook
for
Wetland Identification and Delineation

Practical Handbook
for
Wetland Identification and Delineation

John Grimson Lyon

LEWIS PUBLISHERS
Boca Raton   Ann Arbor   London   Tokyo

Library of Congress Cataloging-in-Publication Data

Lyon, John Grimson.
  Practical handbook for wetland identification and delineation / by John
Grimson Lyon.
      p. cm.
  Includes bibliographical references and index.
  ISBN 0-87371-590-X
  1. Wetlands—United States—Classification. 2. Land use—United
States—Planning. 3. Wetland conservation—United States 4. Wetland
ecology—United States. 5. Wetland flora—United States. I. Title.
QH104.L95   1993
574.5′26325—dc20
                                                                    92-43804
                                                                         CIP

Direct all inquiries to CRC Press, Inc., 2000 Corporate Blvd., N.W., Boca
Raton, Florida 33431.

PRINTED IN THE UNITED STATES OF AMERICA
2 3 4 5 6 7 8 9 0
Printed on acid-free paper

Preface

Identification and delineation of wetlands has recently become an important topic. Wetlands have received new interest from a variety of groups. Along with the recognition of the wetland values has come an increased concern for maintenance of the quantity and variety of wetlands. Laws are now applied in new manners to ensure preservation of these resources. Actions by government regulatory agencies have increased, and the situation is one of controversy and potential change.

This book addresses these concerns and interests by defining wetlands, describing their characteristics, and providing a variety of methods to identify their extent. Chapters examine the soil, hydrology, and plant characteristics of wetlands and the general chemical and biological processes that create these wetland environments. Other chapters detail methods to identify and delineate wetland boundaries, both at the reconnaissance and intermediate levels of detail. Major focuses include additional uses of soil and plant measurement methods, maps, Soil Surveys, aerial photographs, surveying, and other techniques to assist the reader in characterizing wetlands in detail.

Insights are provided on how to optimize wetland delineations and how to produce wetland reports. The use of these suggested techniques and those suggested by wetland delineation manuals will allow the user to complete wetland evaluations. This information will facilitate planning and management of wetlands and adjacent upland resources. The book will be of value to scientists and engineers, landowners, attorneys, regulators, and environmental and conservation groups.

Acknowledgments

Dr. Lyon has been fortunate to have many people assist him in wetland related work, and assist in the development of his career. He would like to thank these people for their assistance over the years. They include: William Aymard, Alfred Beeton, William Benninghof, William Bloch, James Boyle, Bert Brehm, Jack Butterbaugh, Dafney Main Butterbaugh, Deborah Coffey, Marshall Cronyn, David DeSante, Robert DeWall, Jr., Scott Doran, Ronald Drobney, Tom Freitag, Roger Gauthier, Luke George, Raphael Gurvis, Charles Herdendorf, Elise Hoffmann, Dottie Krise, Webster Krise, Terry Logan, Ross Lunetta, James Lucus, Beth Lyon, David Lyon, Lynn Lyon, Ronald Lyon, Sarah Lyon, Ronald Maciak, Jack McCarthy, Gloria Mighell, Robert Mighell, Roger Murphy, James O'Neil, Charles Olson, Jr., David Pyle, John Ray, Jeff Reutter, David Rose, Laurens Rubin, Robert Seedlock, Robert Sierakowski, Douglas Southgate, Otis Sproul, Helen Stafford, James Sturdevant, Robert Sykes, Warren Walker, Andy Ward, Don Wick, David Williams, Don Williams, Dan Williamson, Dorsey Worthy, and T. H. Wu.

About the Author

John Grimson Lyon is an Associate Professor of Civil Engineering and has been a member of the Faculty of Ohio State University since 1981. His interests in wetlands began as a youth and were developed through university education. Dr. Lyon wrote his Bachelor's and Master's Theses, and his Doctoral Dissertation on wetlands. Dr. Lyon was educated at Reed College in Portland, Oregon and at the University of Michigan in Ann Arbor.

Lyon's wetland research has focused on coastal and riverine wetlands, with emphasis on identification and delineation. Much of this work has involved use of aerial photointerpretation, photogrammetry, surveying, remote sensing and geographic information system technologies for characterizations of wetlands. These efforts are documented in the literature cited in the Bibliography.

Dr. Lyon has conducted many wetland delineations and this experience is contained in the book. He has studied wetlands with the US Army Corps of Engineers (USACE) as a guest worker from 1985 to 1988. His wetland and other research has been sponsored by the USACE, NOAA Sea Grant, NASA, NWF and USEPA. He currently serves as an OSU employee on a USEPA cooperative agreement, and functions as a USEPA Visiting Scientist at EPA's Environmental Monitoring Systems Laboratory in Las Vegas, NV. His EPA work focuses on developing satellite data sets of land cover and historical changes in land cover in support of EPA's Global Change Research Program.

Contents

1

Introduction

Identification and delineation of wetlands has become an important topic. This is partly due to the variety of groups that have interest in wetlands. While the viewpoints of each group differ as to the valid uses of these areas, each shares a concern for wetlands. Each group requires considerable information to advance and defend their agenda for the resource. Knowledge of wetland identification and delineation are vital pieces of that information.

The prevailing condition in the U.S. is that wetland laws are applied in new manners as compared to historic applications. The level of enforcement and the varieties of wetlands under scrutiny by governmental regulatory agencies have increased. The advent of new state and federal laws presents opportunities for additional oversight. No net loss of wetlands is the current goal. The future condition is difficult to predict, but change is probably destined to occur.

In this time of concern over wetlands, many groups have compelling reasons to identify and delineate wetlands. The reasons important to an individual group, however, may be very different and as varied as the wetland-related objectives of each group.

Financial institutions frequently require wetland assessments as part of an environmental evaluation. This evaluation and the resulting "environmental report" are often necessary for granting loans for development of property.

Civil engineers need delineations of wetlands prior to site planning. The property site plan must be designed so as not to disturb existing resources.

Conservation, recreation, and environmentally oriented groups may wish to identify especially valuable wetlands for purchase as preserves or parks. For these groups, the primary use of wetland knowledge is for interpretation of the wetland reports of others. With such capabilities, conservationalists and environmentalists can provide learned dialogue on development-related issues.

Local and state governments need wetland assessments for management and planning of existing properties or evaluations of properties for future acquisitions. Governments also need wetland analyses when they expand roads, utilities, and execute other governmental functions. As with the case of any owner of property, governmental agencies can only act with knowledge of the wetland resources that will be impacted.

The above examples are all valid reasons for interest in wetlands and for having knowledge of the methods for identifying their presence and mapping their locations. The motivations of each group or interest are different, yet each needs to have knowledge of wetland resources to advance their respective agendas.

Federal governmental interest in wetlands stems from a number of existing laws (Goldfarb, 1988; Want, 1992). The application of these mandates is currently receiving a higher level of attention.

A 1990 executive order by President Bush expanded the scope of permitting activities with the goal of no net loss of wetlands. This order has resulted in an increase in oversight by the U.S. Army Corps of Engineers (USACE) under historic laws including Section 404 of the Clean Water Act (CWA) and Section 10 of the Rivers and Harbors Act of 1899. Section 404 of the CWA also regulates discharge of sediment-laden waters and materials into wetlands, and it has also received attention under the leadership of state water quality agencies and/or the U.S. Environmental Protection Agency (USEPA). The advent of the Food Security Act of 1985 and Farm Law of 1990 has also focused attention on wetland resources found on farmed lands (Napier, 1990). State laws are very important in each local, and attention must be paid to these requirements.

As the primary federal agency concerned with wetland regulation, the USACE has moved vigorously to administer their mandates. The attention of the USACE to the regulatory element of the wetland issue has resulted in enforcement actions, which were unanticipated by developers, engineers, and contractors. Project delay and work interruption have been and are today real, potential problems for landowners who are ignorant of prevailing wetland laws.

Increased regulation has lead to concerns on the part of the public, and this level of action has contributed greatly to the dialogue on the wetlands issue. The public has become involved by increased activity including the identification of potential illegal fills. Part of this activity results from the wide spread perception that large quantities of wetlands are still being filled. This is thought to occur despite the current level of enforcement by the USACE and the renewed awareness of wetland-related sanctions available to the public through governmental regulators.

In this period of controversy, there is limited information on how to proceed and how to provide an assessment of wetlands suitable for permitting activities of the USACE. In the absence of detailed, governmental information on local jurisdictional wetlands, it has become necessary for individuals to obtain wetland assessments before property is developed (Figure 1.1).

This new level of federal oversight has pointed to the lack of general knowledge related to regulatory requirements and lack of methods documented in the popular literature (Figure 1.2). The goal of this book is to present the characteristics and indicators of wetlands that are the focus of the jurisdictional wetland issue, and present strategies and methods for making wetland identifications and delineations to meet federal requirements. In this period of regulatory change and competing interests, it is hoped that the contents of this book may provide valuable and thoughtful assistance.

Figure 1.1 Isolated wetlands and wetlands in general are hard for the lay person to identify. Pictured is a hardwood swamp in central Ohio.

Figure 1.2 The lay person may be hard pressed to separate wetlands identified in Figure 1.1 from uplands pictured here. The areas are adjacent, but are separated by 6 in. of elevation. Note the Facultative Upland (FACU) plant Mayapple (*Podophyllum peltatum*) dominates the ground layer.

2

Background

WHAT IS A WETLAND?

A wetland can be many things to many people. Generally, the word "wetland" conjures an image of a river or pond surrounded by cattails, alive with ducks, fish, frogs, and the like paddling about (Figure 2.1). Many types of wetlands depart from this model and are often overlooked by people (Figure 2.2). These are the same areas that are often the subject of later, regulatory attention.

There is a need to recognize a variety of wetlands to ensure correct management and to comply with federal and state statues and regulations. To appreciate the variety of wetland types, it is necessary to understand the environmental conditions that foster wetlands and to learn the characteristics and indicators of wetlands.

Wetlands may be hard to distinguish from adjacent terrestrial or aquatic areas. This begs the questions: What is a wetland, and what are the characteristics of wetlands? There are a variety of definitions, just as there are a variety of people with ideas of "model" wetland types.

A wetland can be described as a mix of characteristics from terrestrial or upland areas and the characteristics of aquatic or water environments. In essence, a wetland is the edge or interphase between uplands and adjacent water areas. The water may be in the form of rivers, lakes, ocean areas, or wet spots. As such, wetlands may be found almost anywhere. They will possess characteristics of both upland and aquatic environments and exhibit a mix of soil, plant, and hydrological conditions (Figure 2.3).

Figure 2.1   Riverine wetlands in the Scioto River, Upper Arlington, Ohio. Note the elements of a common definition of wetlands such as high water, wetlands plants, and wildlife.

Figure 2.2 Great Lakes coastal wetlands including the nearshore high energy beach, beach barrier wetlands, and adjacent washover fans. The location is the Straits of Mackinac of Lake Michigan, low peninsula of Michigan. This is also nesting habitat for the endangered Piping Plover.

Figure 2.3 The combination of hydric soils, hydrophytic plants, and wetland hydrology are evident here. The same area was waterless
a fortnight later following evaporation processes. This ephemeral wetland was found in Valley of Fire State Park, southern
Nevada.

This mix of characteristics creates a unique habitat for life and earth processes, but the mix also makes wetlands hard to identify. This is due to the inherent gradation of these characteristics from uplands to the aquatic environment and the presence of both kinds of conditions in various combinations along that gradient.

There are a variety of formal wetland definitions. Several have been used by federal agencies. Four federal agencies have collaborated over a number of years to arrive at an "accepted" definition for use in their activities. This particular definition is most important because elements of the definition are used administratively by the U.S. Army Corps of Engineers (USACE) for evaluations of wetlands. It is also important because it embodies three criterion for identifying a wetland and thus provides three measures to judge whether an area is truly a jurisdictional wetland. It is also important to recognize that this definition may change in the future.

The definition comes from the interagency, "Federal Manual for Identifying and Delineating Jurisdictional Wetlands." Here, it is referred to as the "Federal Manual" or FICWD (1989). Most of the definition stems from the USACE regulations, and previous efforts to define wetland characterists.

According to the Federal Manual, a wetland can be defined as:

> Those areas that are inundated or saturated by surface or groundwater at a frequency and duration sufficient to support, and that under normal circumstances do support, a prevalence of vegetation typically adapted for life in saturated soil conditions. Wetlands generally include swamps, marshes, bogs, and similar areas. (EPA, 40 CFR 230.3 and CE, 33 CFR 328.3)

This definition is used by the USACE, U.S. Environmental Protection Agency (USEPA), and other agencies and by people using the Federal Manual (FICWD, 1989) or the USACE (1987) Wetland Delineation Manual for their work. In practice, it is an operational definition. From it stems the three criterion used as the standards to be applied in identifying jurisdictional wetlands (Figure 2.4).

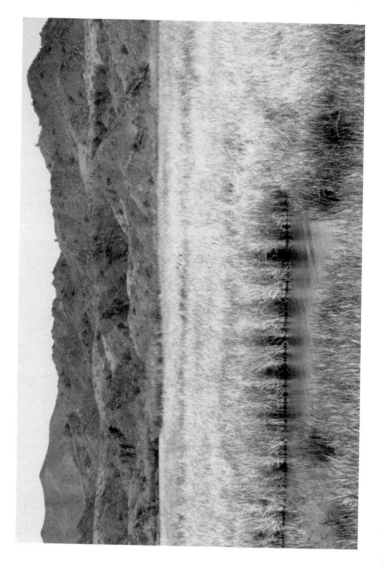

Figure 2.4 Note the contrast between desert hills surrounding the Colorado River and its Lake Havasu reservoir, and the emergent wetlands on the reservoir margin. The wetlands are found at the terminus of the Bill Williams River on the Arizona side of Lake Havasu.

The waters addressed in the application of this definition include those important to the USACE. Under Section 404 of the Clean Water Act (CWA), this definition may be applied to regulate discharges of dredge or fill material in the waters of the U.S. These areas include navigable waterways, most lakes, rivers, streams, impoundments, wetlands, sloughs, prairie potholes, wet meadows, ponds, and the like that have interstate or foreign commerce connection (USACE, 1992).

This definition and the application of it to waters under the USACE jurisdiction results in most wetland areas falling into the domain of regulatory oversight. Other wetlands found above the headwaters and found to have indicators of all three wetland criterion are also subject to the regulatory attention of the USACE when discharge or filling is to occur.

There are several other definitions of wetlands that may be important. This is particularly true of wetland areas that may be farmed lands or areas where state or other laws based on other definitions also prevail. There are also important historical definitions that may be or should be taken into account based on local considerations (Shaw and Fredine, 1956).

The use of a given definition depends on the characteristics of the study area and interests of the particular agencies or groups involved in wetland determinations.

An important definition is that used in the Food Security Act of 1985 because it is applied to farmed lands:

> Wetlands are defined as areas that have a predominance of hydric soils and that are inundated or saturated by surface or ground water at frequency and duration sufficient to support, and under normal circumstances do support, a prevalence of hydrophytic vegetation typically adapted for life in saturated soil conditions, except lands in Alaska identified as having a high potential for agricultural development and a predominance of permafrost soils (National Food Security Act Manual, USDA, 1988).

The whole issue of agricultural practices and wetlands is an evolving one, and it presents an incredible uncertainty for agricultural interests.

The Emergency Wetlands Resources Act of 1988 contained the same definition, but the reference to Alaskan lands was deleted (Figure 2.5).

Figure 2.5 Is this a jurisdictional wetland? Is this a wetland for purposes of the National Food Security Act? Probably not, due to the ephemeral nature of desert rains and ponded water. Pictured is a playa lake area south of Boulder City, Nevada.

The U.S. Fish and Wildlife Service has also developed a definition of wetlands used in the National Wetlands Inventory Program or NWI Program. The definition states:

> Wetlands are lands transitional between terrestrial and aquatic systems where the water table is usually at or near the surface or the land is covered by shallow water. For purposes of this classification wetlands must have one or more of the following three attributes: (1) at least periodically, the land supports predominantly hydrophytes, (2) the substrate is predominantly undrained hydric soil, and (3) the substrate is nonsoil and is saturated with water or covered by shallow water at some time during the growing season of each year (Cowardin et al., 1979).

THE FEDERAL DEFINITION AND CRITERIA FOR IDENTIFICATION OF WETLANDS

The finding of a jurisdictional wetland is based on criteria set forth in the Federal Manual (FICWD, 1989) or the manual of current application such as the 1987 USACE Wetland Delineation Manual (USACE, 1987). An area is considered a jurisdictional wetland only if all three wetland criterion are met. These requirements may also change in the future.

The evaluation of these criterion includes a determination as to: (1) whether the soils are considered hydric or waterlogged, (2) whether the soils show demonstrable evidence of hydrologic conditions associated with flooding or ponding of water, and (3) whether 50% of the dominant plants found growing on the site are those commonly found in wetlands.

Fundamentally, property areas that fail to satisfy one of the three wetland criterion are not considered jurisdictional wetlands. The exact conditions depend on the delineation manual that prevails at a given moment, and one should be aware of current conditions based on the manual and any prevailing regulatory guidance letters from the USACE. State laws must also be taken into consideration.

Properties or wetlands that have been disturbed provide exceptions to the requirement that three criterion be satisfied. These areas may not exhibit one or more of the criterion, due to burial or removal in years past when the public was unaware of the value of wetlands. This altered condition represents a special case, and it is much different than an illegal fill situation.

These disturbed areas are evaluated using less than the total, three criterion. The procedures are similar to a regular evaluation, except that the wetland elements that are disturbed or missing are no longer a factor in the analyses. This procedure may be used on illegal fills, but the legal and permitting consequences are much different. This procedure is described in the Federal Manual (FICWD, 1989).

It is desirable, here, to address the three criterion in detail. This approach allows for appreciation of phenomena that cause an area to exhibit wetland conditions. It also demonstrates how information on hydric soils, wetland hydrology, and wetland plants helps to define a jurisdictional wetland area. It can also promote the understanding of indicators employed in an assessment of jurisdictional wetlands, and it can help one appreciate the complex mix of the three characteristics that make a wetland.

Hydric Soils

If a given soil is subject to flooding or ponding of water for more than one to two weeks per year, it will often demonstrate hydric or waterlogged soil characteristics. These waterlogged conditions greatly influence soil chemistry and the conditions for plant life (Figure 2.6).

The significance of standing water or waterlogged soils is that chemical and biological oxygen demands will rapidly exhaust oxygen available in the soil. This is significant because a lack of soil oxygen for roots will commonly cause death of upland plants during the growing season. All plant roots need air to respire and use airborne oxygen to metabolize sugar and supply energy for life.

Diffusion of oxygen into soil is a slow process, in general, and very slow into water as compared to movement or diffusion in the air. Migration of oxygen into waterlogged soils and into the root zone of plants is also very slow and proceeds at an unacceptable rate compared to the oxygen needs of upland plants and other biological and chemical oxygen demands in the soil.

Figure 2.6 Hydric soils underlie this Great Lake coastal wetland in Saginaw Bay, Lake Huron, Michigan.

Hence, areas that experience periodic flooding or ponding of water are often populated by plants adapted to poor oxygen conditions in the root zone. These plants are commonly known as wetland plants (Figure 2.7).

Soils that experience these oxygen-poor or anaerobic conditions on a periodic basis are characterized as being hydric. The U.S. Department of Agriculture's (USDA) National Technical Committee for Hydric Soils Criteria has developed a list of soils that often display hydric soil characteristics.

Potential hydric soils can be identified by examining a USDA county Soil Survey. Soil types found on the site of interest can be compared to those on the Hydric Soils List. Presence of hydric soils on a site is often interpreted administratively by federal agencies and other groups as being indicative of potential wetland soil conditions and potential jurisdictional wetlands.

Evaluation of the county Soil Survey for soils on the USDA Hydric Soils List (USDA, 1991) is a necessary first step in an evaluation of a given property. If "Hydric List" soils are found on the property, it will often be incumbent upon the landowner to determine if jurisdictional wetlands are present by analysis of the site using the three criterion of the Federal Manual or another current wetland delineation manual.

In a practical sense, the presence of soils from the USDA Hydric Soils List indicates a potential jurisdictional wetland condition. This means that the property may be subject to administrative review by the USACE for the presence of jurisdictional wetlands should the landowner propose to change the end-use of the property. This is because the Soil Survey and Hydric Soils List are often used by federal agencies and others as a "first cut" evaluation of the potential for wetlands.

Soil Survey information and other sources such as NWI maps may be interpreted by the USACE personnel in judging the probability of finding jurisdictional wetlands on a given site. The delineator should pay attention to those characteristics of a given site and take the condition into account when making wetland determinations.

Soil Surveys are available from the county offices of the Soil Conservation Service (SCS) of the USDA. These offices are commonly found in the county seat and can be identified in

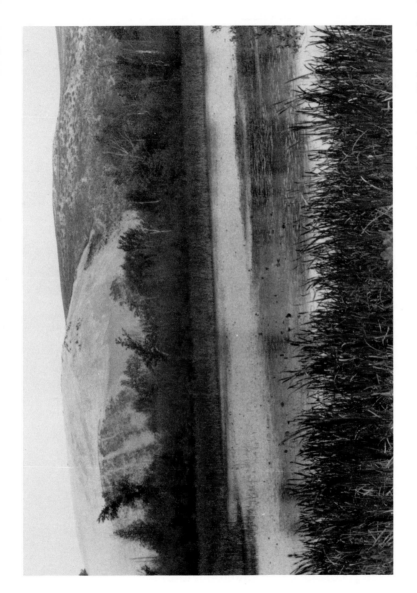

Figure 2.7 A variety of wetland plants fill and border this lake. Included are cattail (*Typha*), bulrushes (*Scirpus*), rushes (*Juncus*), sedges (*Carex*), birches (*Betula*), and willow (*Salix*) species. The background is Sleeping Bear Dune on Lake Michigan in Leelanau County, Michigan.

the governmental section of local telephone directories or through directory services.

The USDA Hydric Soils List can be obtained from the local SCS office or from the state headquarters of SCS in each state capitol. The list is also available from the USDA-SCS in Washington, DC, National Technical Committee for Hydric Soils, Criteria for Hydric Soils (NTCHS, USDA, SCS, 1991). It is also available from USDA-SCS, Soil Survey Division, POB 2890, Washington, DC, 20013 or from NTCHS, SCS, Room 152, Federal Building, 100 Centennial Mall North, Lincoln, NE, 68508-3866.

Hydric soil indicators can be identified in the field. This generally involves observations made by digging a hole or probing the soil to approximately 18 in. in depth. The soil color is evaluated using hue, value, and chroma characteristics from the Munsell Color Chart for soils (Figure 2.8). Observations of the macroscopic manifestations of anaerobic soil chemistry should also be made and can be used in identifying water-logged conditions.

Wetland Hydrology

Soils and plants that are generally flooded display a number of characteristics diagnostic of waterlogging or anaerobic soil conditions (Figure 2.9). These characteristics or indicators include the presence of soil colors or mottling, dark soil color or chroma, as well as other indicators including the presence of chemical deposits associated with chemical reactions that occur in the absence of oxygen.

Other overt signs of influence of water and conditions of wetland hydrology include flood "markings" on the soil surface and coatings of clay or silt particles which may appear on plants. Rafted debris may be present on the ground at the margin of flooding (Figure 2.10). Materials or debris may be found suspended in trees several feet above the ground.

During inspection of the soil hole, it is desirable to observe whether water seepage is encountered within 18 in. or so of the soil surface. This is a strong indicator of wetland hydrological conditions and an important jurisdictional measure of soil hydrology. An obvious filling of the hole with water is a very good indicator of a seasonal high water table. It can be readily

Figure 2.8 Field personnel check for soil conditions in a soil hole or pit. The Munsell Color Chart is read for soil matrix colors of each horizon that can be observed. Personnel also check for the presence of mottling and other hydric soil indicators, and seepage or drainage conditions of wetland hydrology. Mottled colors or high anaerobic gleyed soils are then described with the Munsell Color Chart.

Figure 2.9 The variation in reservoir lake level and local elevational difference influences the growth of wetlands in Lake Havasu, Arizona.

Figure 2.10   Conifer needles form duff on the floor of an ephemeral stream. The location is rural Georgia.

interpreted as meeting the hydrological criterion based on the prevailing manual and its criteria.

Another method of determining potential wetland hydrological conditions is to evaluate the proximity of a given property to rivers, streams, gullies, and nearby wetlands of significance. This can be performed without a field visit and may be completed by using available maps or aerial photos. Federal Emergency Management Agency (FEMA) flood plain maps are also useful for identifying adjacent riverine characteristics and may be obtained from FEMA (Appendix A).

All these characteristics can be important in the assessment of jurisdictional wetlands. These are all descriptive indicators, or simple measurement techniques, and may be infrequently used in some areas of science or engineering. However, the Federal Manual or other manuals specifically ask for these sort of data, and it is best to focus data collection on the required information. Certainly more detailed measurements can be employed if available or if their use is requested by the USACE.

Wetland Plants

The third important criterion for wetland determinations is whether the site has an abundance of plants adapted to grow in wetland or waterlogged soils. In general, plants are adapted to grow in specific environments and compete with each other for nutrients and light. One competitive advantage for a given plant species is the capability to grow and reproduce in the air- or oxygen-free environment of waterlogged or anaerobic soils.

Waterlogged soils present a very stressful environment for many plants. However, for the plant species that have capabilities to grow in this environment, it can provide a competitive advantage over other plants and provide quality habitat for the wetland plant. Wetland-adapted plants can grow in exclusion of upland plants, which can only populate air or oxygen-rich environments. Wetland plants exploit this stressful environment to grow and reproduce and thereby "out-compete" upland plants for a habitat or "home."

There are a number of plants capable of growth and reproduction in wetlands. These plants exhibit structural or physiological adaptations to accommodate their growth in wetland areas. The adaptations include plant biochemistry optimized for anaerobic soil conditions and the presence of plant tissue for conduction of air from above to the roots (Sculthorpe, 1967; Lee et al., 1975; Good et al., 1978). Plants may also change their form of growth to maximize surface contact with the air environment through structures such as shallow roots, multiple tree trunks, or roots on the surface of the ground (Figure 2.11). Plants that have these characteristics are commonly found in wetland environments (Sculthorpe, 1967; Hutchinson, 1975; Teskey and Hinckley, 1978).

The identity of plant species that can exist in wetlands is known and generally agreed upon within the scientific community (Reed, 1988; Gosselink et al., 1990). A list of plant species and their affinity for wetland conditions has been published and is referred to here as the "National List" (Reed, 1988). The full title of the list is the "National List of Plant Species That Occur in Wetlands: 1988 National Summary," published by the U.S. Fish and Wildlife Service (USFWS) of the Department of Interior.

The National List is a fundamental resource for identifying the wetland affinity of individual plant species. This book is used in wetland determinations for jurisdictional purposes. It is important for all interested parties to obtain a copy or have access to a copy of Reed (1988) for reference purposes. Sources for copies include the U.S. Government Printing Office, the National Technical Information Service (NTIS), and libraries (Appendix A).

The capability of a given plant to live in a wetland and the probability of it being found in a wetland are rated in one of five categories. This assessment has been made for 13 different regions of the U.S.

The system of categories (Reed, 1988) recognizes that certain plants are found almost exclusively in upland environments (system category: Uplands or UPL). A number of plant species may be found in wet environments less than 33% of the time and are termed facultative upland plants (category: FACU). Some species usually occur in wetlands with an estimated probability of 34 to 66% of the time, but are occasionally found in nonwetlands and are called facultative (FAC) wetland species. Some plants are usually found in wetlands with an estimated probability of 67 to 99% and are called facultative wetland (FACW) plants (Figure 2.12). Certain plants occur almost always in wetlands with an estimated probability of >99% and are called obligatory (OBL) wetland plants (Figure 2.13). Often times a suffix of "+" or "–" is applied to indicate the plant is found on the higher or lower ends of the range of probabilities.

In a wetland delineation, the plant species that are the most common or dominant must be identified within a given area. These common plants must be estimated as to their relative abundance (Figure 2.14). One simple method is to estimate dominance for each common plant species on an areal extent or coverage basis in percent. Such a determination may be made for each dominant plant species in each layer of vegetation including the tree, shrub, and ground or herbaceous layers.

It seems to be the best approach to separate the vegetative cover into the common three layers of tree, shrub, and ground. If a vine or other demonstrable layer or stratum of vegetation exists in ad-dition to these mentioned here, it may be used in the analysis.

The assessment of vegetative dominance of wetland plants incorporates the sum dominance of the entire sample, and that is the deciding factor. However, the measurements may be split up by layer. However, they must all sum to the total, and the number of divisions of the whole is immaterial.

The expert determines the status of wetlands in a given area by recording all dominant plants. The determination itself involves enumerating only the plants most capable of existing in wetlands. These are the dominant plants in the FAC, FACW, and OBL categories of plants (Appendix B). This is in accordance with the current manuals.

A given site is determined to be a wetland for this criterion when the total dominance of FAC, FACW, and OBL plants exceeds 50% of the total dominant plants found on the site.

There are a variety of methods recorded in the scientific literature for measurements of plants (see Bibliography and manuals including USEPA, 1991). There are a number of methods favored by individuals and a number of methods that are

Figure 2.11 Note the spreading tree roots above ground. This is an adaption to seasonal high water table by roots of this tree in an isolated hardwood swamp in central Ohio. This "big base" tree is an indicator of wetland hydrology, along with the poorly decomposed leaf litter above the ground.

Figure 2.12   Facultative Wetland (FACW) plants in a cedar swamp, upper peninsula of Michigan.

Figure 2.13   Obligatory (OBL) wetland plants in a pond near Stone Mountain, Georgia.

Figure 2.14   Estimation of relative abundance is important in establishing a list of dominant plants. Note the complex mix of species including *Scirpus, Juncus,* and *Carex* in this emergent wetland in Leelanau County, Michigan.

workable and can be employed in analyses. For delineations, it is both necessary to estimate the dominance and relative abundance of dominant species, and it is necessary to do so using methods not overly time consuming.

Determination of dominance may be completed by visual estimation or by more complex estimation procedures such as the point intercept method or some quadrant based methods (Mueller-Dombois and Ellenburg, 1974; FICWD, 1989). The method of choice is up to the person conducting the delineation, and the approach should be acceptable to the USACE. The method should be repeatable and uniform in its application.

3

Methods

A number of methods can be used to delineate wetland areas. As per the Federal Manual (FICWD, 1989) and other manuals, potential jurisdictional wetlands must be identified by addressing the three criterion. This may be accomplished by using a combination of data sources and field indicators. Different levels of attention may be applied based on the need for information and documentation.

To identify and delineate wetlands in a given area, it is best to use a combination of existing data, reference materials, and field evaluation procedures. This approach allows all three criterion to be addressed in detail and their conditions recorded. Field measurements may involve ground examinations of hydric soils identified on Soil Survey maps, evaluations of waterlogged soil indicators and/or flooding associated with wetland hydrology, and estimations of dominant plant species and the relative percentage of those plants commonly found in wetlands.

The following activities have been found to be beneficial for identification of jurisdictional wetlands and for delineating and mapping wetlands.

SELECTING A METHOD OF ANALYSIS AND LEVEL OF DETAIL

An important early consideration in any project is the level of detail and effort necessary to identify and delineate wetlands. The Federal Manual (FICWD, 1989) and other manuals present three levels of effort that yield an increasing quantity of detail. Selection of the level of effort can be very helpful in tailoring the work product to the need for information.

The routine or reconnaissance level evaluation involves office efforts and perhaps minimal field work. It is a very good starting point, as it reveals the general extent of wetlands and identifies the potential jurisdictional wetland areas.

The intermediate level evaluation employs field-based methods to evaluate wetland resources, to quantify wetland characteristics, and to produce a detailed report. The results of this delineation and the intermediate level report present a good level of information and can be used in permitting by the U.S. Army Corps of Engineers (USACE).

Comprehensive level methods are very detailed and laborious. This level of delineation may be required for special case wetland areas under intense scrutiny or for better characterizing wetlands for purposes of wetland function evaluations in support of mitigation. This level may also be appropriate when a group wants to characterize the wetland habitat with greater detail and facilitate their management of wetland and other resources.

Routine Level Analysis Procedures

The preliminary or "routine" level of analysis is a very good place to start any effort. This is because a "reconnaissance" can be conducted to characterize the scope of the problem. The routine evaluation indicates where potential jurisdictional wetlands may be located. It would involve an office evaluation of the Soil Survey and other sources of information, and a field inspection for the presence of wetland plants, hydric soils, and wetland hydrology. Further detail can be collected within this framework by later applying intermediate level evaluations to determine whether local areas are truly jurisdictional wetlands.

For the purposes of routine level analyses, there is a need to define the concept of a "potential jurisdictional wetland." This label or category allows for documenting a potential condition before the evaluation of all three wetland criterion has been made. Usually, it is necessary to "scope" the potential jurisdictional wetlands. This scoping or routine level wetland report will allow the client to appreciate the conditions and choose whether to go forward with a more detailed, intermediate level effort.

Hence, it is desirable to define the concept of a "potential jurisdictional wetland" as an area that exhibits one or more of the three wetland criterion. Wetlands identified during routine level evaluations may be jurisdictional wetlands, but it is necessary to complete an intermediate level evaluation to make an actual determination as to whether the area is a jurisdictional wetland. The intermediate level wetland determination would involve greater field sampling with increased detail on plants, soils, and hydrological conditions.

To complete a routine investigation and identify potential jurisdictional wetlands, it is desirable to use the following suggested steps and procedures:

1. Obtain the county Soil Survey, and transfer the soil boundaries from Soil Survey maps onto a large scale topographic map or USGS quadrangle map of the site. This should be done for all soil types on the property with particular attention to soils found on the USDA Hydric List (USDA, 1991).

2. Take the soil maps and other data sources such as aerial photos, National Wetlands Inventory (NWI) maps, and large scale topographic maps to the field, and walk the site. Pay attention to lower elevational areas, all streams, creeks, gullies, wet spots, and wetlands. Evaluate any anomalous conditions of soil, bedrock, or hydrology that could result in flooding, ponded water, or a high water table.

   Walk the entire site alert to these and other conditions, and make notes as to the presence of indicators of hydric soils, wetland hydrology, and wetland plants.

   Evaluate the entire site, and avoid omission of small or hidden wetlands. This is vital in assuring that all potential jurisdictional wetlands are identified and mapped. It is also important because the USACE will often wish to inspect the entire site, and omission of wetlands can create many difficulties.

   Pay particular attention to hydric soil areas as mapped by the Soil Survey. Also, look for wetland plants that may "signal" the presence of hydric soil conditions that do not appear on the Soil Survey, or conditions such as hydric soil inclusions within the larger Soil Survey mapping unit.

Note the presence of wetland plants, in particular the obligate wetland plants and facultative wetland plants.

3. Take the field information, available aerial photographs, the county Soil Survey, other pertinent data, and identify potential wetland areas. Any area with one or more wetland indicators should be flagged or marked as a potential jurisdictional wetland. These locations can be reported and later sampled at the intermediate level of detail to determine if true jurisdictional wetland conditions are present.

Map the boundaries of potential jurisdictional wetlands on an overlay of the property map or large scale topographical map. The desirable minimum mapping unit should be $1/10$ or $1/100$ of an acre depending on the scale of available maps. Using a planimeter or another area measurement device make a preliminary estimate of the total potential jurisdictional wetlands.

4. Integrate information from other data sources such as watershed and drainage details from topographic maps to help characterize any additional resource conditions that may be of jurisdictional interest.

5. Check the routine wetland report. This may include an additional field visit. It is important to be sure that no potential jurisdictional wetlands have gone unnoticed. The landowner is depending on the expert for an accurate assessment of potential jurisdictional wetlands. Later, the landowners will conduct their efforts based on the report. An inaccurate report can cause incredible difficulties after commencement of development should the USACE become involved and a conflict result as to the quantity of wetlands on the site.

Completion of these steps provides a certain minimum level of information and allows identification of areas that are potential jurisdictional wetlands. This preliminary or routine level report is suitable for identifying the scope of the problem and is the starting point for further, intermediate level analyses to characterize potential jurisdictional wetlands as jurisdictional wetlands.

Intermediate Level Analysis Procedures

A higher level of detail is supplied by intermediate level evaluation procedures. The resulting product is valuable for making an actual determination of jurisdictional wetland quantities. Such an analysis and the documents resulting from it can be used to respond to wetland related questions posed by the USACE and the Section 404 permitting process.

To complete an intermediate level evaluation, it is necessary to conduct a more intensive effort in the field and office as compared to the routine level of analysis.

These steps are outlined in the Federal Manual (FICWD, 1989) and other manuals and are provided here along with enhancements developed from many applications of these procedures by the author. It is assumed that a routine level evaluation has been conducted to characterize the scope of the problem and that potential jurisdictional wetlands need to be identified and delineated as jurisdictional wetlands.

The steps in the evaluation may include the following:

1. Take data source materials developed from the routine level evaluation, and go to the field. Begin sampling for the three criterion at the location of the potential jurisdictional wetlands.

    For the intermediate level effort, it is desirable to locate the field sampling in some reference framework. This can be accomplished by the establishment of a grid system of points to be sampled in the field. A 100 × 100 ft grid mesh has been found to be of suitable size and resolution for sampling at the intermediate level of detail.

    Such a mesh of sampling locations can be measured by walking or by pacing in the field. Initially, lay out a 100-ft length on the ground, and determine the average number of paces required to cover 100 ft of distance. This "calibration" should be performed by each individual undertaking the wetland study. Repeat this calibration procedure several times to determine the average number of steps per 100 ft of distance. Afterward, lay out 100-ft lengths in the field on a reliable basis by pacing.

    With a set grid mesh and grid size, and directional references for the grid, it is possible to sample at an appropriate frequency and document the location of the samples. The grid can be referenced to cardinal directions, fence lines, or other reference points via compass. It will be possible to relocate the sample sites for later inspection by the USACE personnel.

Be sure to "flag" or monument the individual sample points so that they can be checked at later field inspections (Figure 3.1). Surveyors tape, plastic tape, or other permanent flags should be employed. Time and weather are hard on these materials and care should be taken in selecting materials that will last. Often, it may be many months before a given property is inspected, or it may never be necessary to inspect a property. It is still important, however, to be able to relocate the sampling sites at a later date. A request for an inspection by the USACE may be encountered at any time, and the elements of the field sample should be preserved to facilitate a response to questions posed by the USACE.

It is also desirable to pace from a given sample point to adjacent examples and evaluate whether the grid system is square. This allows one to determine if individual points are out of position, and it facilitates correction of misplaced samples.

In certain instances, it is desirable to use survey methods to establish the grid. Suitable techniques are discussed later.

At each individual sample point or grid-node, it is necessary to collect detailed information on soils, hydrology, and plants. Data should be recorded on field sheets or memo pads. The data can be presented later on the appropriate forms supplied in the Federal Manual (FICWD, 1989). In particular, forms "B-2" and "B-6" have been useful, and they provide adequate space for the recording of plant, soil, and hydrological measures on one form (Appendix B).

2. To evaluate wetland hydrology and subsurface conditions, dig a hole with a shovel at the grid sampling site. The sampling hole should be deep enough to evaluate the top 18 or so inches of soil. It may be sufficient to use a soil probe to evaluate soils, but it is much more difficult to observe soil characteristics mentioned below. In addition, a shovel hole or pit will remain for many months, and the location of a given sample can be inspected by regulatory personnel at a later date.

Describe the soil type found at each sample point. This could include general soil textural conditions, soil levels or horizons, and soil colors. Record whether sample points have similar or dissimilar soils to those indicated by the county Soil Survey.

Figure 3.1 Pictured is a sampling site at a node location of the sampling grid. Note the presence of an excavated hole or pit in the soil for observations of hydric soils and wetland hydrology conditions. The sample location is "flagged" nearby to make it easy to identify the site at another time.

These soil evaluations should include checking the Munsell Color Charts (Munsell Color, 1990) for color, value, and chroma (Figure 3.2). These soil evaluations should also check for iron oxide mottling, oxidized root zones, presence and depth of organic matter and organic soils, presence of manganese reduction products, and/or grey or "gley" deposits.

Be sure to sample each soil layer encountered and record your findings. These layers are commonly called the "A" horizon or top soils, "B" horizon or subsoils, and the "C" horizon or parent material (USDA, 1962; Foth, 1990).

3.   Observe the hole for standing water or seepage of water into the hole from the bottom or from the sides. Observe the conditions for 15 to 20 min after the hole is dug. This can be accomplished by checking the hole after a certain time period has elapsed, while one is conducting other evaluations at sample sites nearby. Also, check for "gley" or very wet soils resulting from very anaerobic conditions, sulfur or methane smell, and other indicators of wetland hydrology.

For jurisdictional purposes, soils that exhibit more than one of these indicators such as standing water or seepage, dark Munsell chromas (/2, /1), lots of mottling, thick layers of high organic matter soils, gleyed soils, or other products from highly anaerobic conditions will be considered to exhibit evidence of the hydric criterion (USACE, 1987; FICWD, 1989).

4.   Check for indicators of hydric soils such as seasonal high water conditions. Such conditions are usually defined as 7 to 14 days duration of flooding or high water table per year during the growing season. In other manuals or proposed manuals, these periods may vary between 15 days of flooding or 21 days of very high water table. These measures of duration of flooding may change in the future so be aware of current requirements.

Record whether the soil and surrounding landscape exhibit hydrological indicators. At each sample point, note flood markings or rafted debris lines, shallow root systems, wet and/or poorly decomposed plant materials, adjacent stream courses, and other indicators of wetland hydrology mentioned in the Federal Manual (FICWD, 1989). In particular, look for a surface layer of undecomposed leaves

Figure 3.2   The Munsell Color Chart is used in describing the soil matrix color of each soil horizon and any mottlings or other soil color conditions.

and/or an absence of plant growth as compared to adjacent areas, or silt and clay deposits on leaves and tree trunks.

5.   Evaluate the plants found at the sample site (Figure 3.3). The ground, shrub, and tree layers of the vegetation need to be described by the plant species that are dominant. If a vine or other vegetation layer or strata are present and plants within them display dominance, they can also be evaluated.

It has been found to be appropriate to evaluate the ground layer of vegetation within a circle of radius of 10 ft centered at the sample point. The shrub layer should be described within a 20-ft radius, and the tree layer within a 30-ft radius circle about the sample point. These radii have proven useful in wetland delineations and are based on suggestions in the Federal Manual (FICWD, 1989) and in other sources. This approach may not be suitable if there are changes in future manuals, so be aware of current requirements.

It is also necessary to estimate relative abundance of each dominant wetland plant species based on the relative prevalence of the given species. This is then compared to the relative percent abundance or cover of the total quantity of vegetation at each sample point.

This may be done by visual estimates, particularly if the observer is experienced in this method of estimation. The Federal Manual (FICWD, 1989) and other manuals provide additional dominance estimating procedures, but the visual estimate is both expedient and accurately performed with practice.

6.   The plant species data and the dominance estimates in percentage are used to identify the common plants at the sampling site. The wetland determination procedure is presented in the Federal Manual (FICWD, 1989) and other manuals (USACE, 1987), and the same or similar techniques are addressed in books or journals (e.g., Mueller-Dombois and Ellenburg, 1974). The procedure consists of determining whether 50% of the dominant plants have a high probability of occurring in wetlands.

The National List (Reed, 1988) records the plant species and their "agreed upon" categories of probability of occurrence in wetlands. Wetland plant species do vary in their probability of occurrence from one region to another, and

Figure 3.3 In the foreground, the wetland plant species are found mostly in the ground layer, and the dominance should be described for that layer. The cedar swamp trees in the background would require a description of plant dominance in the ground, shrub, and tree layers of the vegetation community. These are back-barrier coastal wetlands of Wilderness State Park, Straits of Mackinac on Lake Michigan.

hence the probabilities are listed by the geographical region of interest. Use the intermediate level recording sheets to record and sum the percent coverage of dominants listed as facultative (FAC), facultative wetland (FACW), and obligatory (OBL) types.

7.   Plot the locations of the sampled points on an available topographic map, and supply the boundaries of any jurisdictional wetlands. It may also be desirable, though not necessary, to present the boundaries of any hydric soils encountered. These boundaries will make a nice contrast with any USDA hydric soil boundaries shown on the soil survey map.

8.   To make a finding that an area is a jurisdictional wetland, it is necessary to have: i) hydric soils, ii) evidence of wetland hydrology, and iii) 50% or more of the dominant plants with a high probability of occurring in a wetland (Figure 3.4).

   All three of these criterion must be satisfied. Failure to meet one or more of the three criterion means that the individual location is not considered a wetland for jurisdictional purposes (Figure 3.5).

9.   Use other sources of data and field checks to determine the adequacy and accuracy of mapped jurisdictional wetland boundaries. Adjust boundaries as necessary, and determine the acreage of the actual jurisdictional wetlands. Supply the acreage estimate in the wetland assessment report.

   It may be necessary to revisit the field to locate the exact jurisdictional boundaries as they are often found between the sample points. In essence, it is necessary to interpolate between the sample grid points to fix the boundary. If the characteristics of the wetland boundary are not clear, sample soils and plants to determine the actual boundary as it is found between the sampling points of the 100-ft grid.

   It is also desirable to check and pace the location and size of the wetland, and assure that its characteristics have been correctly mapped. Double check the dimensions of the wetland by pacing to assure the area estimate is correct. It may also be necessary to flag the actual boundaries of the wetland on the ground to facilitate surveying of boundaries or to facilitate inspections by the USACE.

10. Allow a period for evaluation of products by the client. Incorporate comments and criticisms, and produce final products identifying jurisdictional wetland areas.

Figure 3.4 Is this temporary drainage ditch a jurisdictional wetland? The area displays wetland hydrological indicators and saturated soils. The soils are mottled and hydric. The dominant cattails and some other smaller plants are wetland species (either OBL, FACW, or FAC). Hence, all three indicators are present, and the area is a jurisdictional wetland.

Figure 3.5   Is this small wet spot a jurisdictional wetlands? All three indicators are present, and hence it is a jurisdictional wetland. The spot is approximately 15 ft wide and has approximately 4 to 6 in. of peaty, wet soil underlined by the granite on top of Stone Mountain, Georgia.

At this point, the wetland report may be filed, or the information used in planning and management of the wetland resource. A sample intermediate level report is provided in Appendix B.

These products and a report summarizing the methods used in their production may be submitted as a wetland assessment to the USACE in support of Section 404 permitting activities.

The information and methods provided above provide some insight and appreciation of the complexities involved in wetland evaluations. Intermediate level methods can be useful in a number of applications, and they provide a good, general model of how to acquire and present data for determination of jurisdictional wetlands.

The intermediate level report is very suitable for permitting activities and represents the minimum level of detail necessary to support a finding of jurisdictional wetlands. It is particularly desirable to document the preconstruction conditions of a given property using these techniques, even if no jurisdictional wetlands are found. Completion of an intermediate report provides very good documentation if questions arise after initiation of construction. It is good evidence of due diligence on the part of the landowner in fulfilling responsibilities related to wetlands.

Comprehensive Level Analysis Procedures

In certain situations, it may be necessary to use additional techniques to augment those presented above or described in the Federal Manual (FICWD, 1989) or other manuals. The need for comprehensive procedures to supply great detail may be manifested by the complexity of the wetland, the complexity of the terrestrial and aquatic interface, or the complexity of the proposed new constructed wetland environment.

Detailed, comprehensive level procedures may be necessary because the status and extent of the wetlands are in contention. Other needs may be dictated by the presence of a unique plant or animal species that uses the wetland as a habitat or "home" (Figure 3.6), and wetland characteristics must be detailed for more than the needs of identifying a jurisdictional wetland and Section 404 permitting activities.

Figure 3.6 The continuum from upland or terrestrial habitat conditions to wetland conditions and to aquatic conditions of the open pond is displayed in this rural Georgia scene.

The following chapters supply a number of methods that may be used to augment the intermediate level procedures. Some of these methods are defined as comprehensive-type methods in the Federal Manual (FICWD, 1989) and other manuals (USACE, 1989; USEPA, 1991). Other approaches are suggested by the author based on personal experience.

The background information provides a higher level of detail and techniques supplied in the following chapters can be used to customize the wetland evaluation process to meet the unique characteristics of a particular site.

4

Additional Background and Details

SOILS

Hydric soil characteristics need to be understood and appreciated to make a good identification and delineation of wetlands. Soil conditions need to be evaluated using methods from the Federal Manual and other manuals and sources. These sources will help to adequately address the hydric soil criterion in determination of jurisdictional wetlands.

Wetland soils are formed wherever water stands for a period of time or where there is a frequent presence of water (Figure 4.1). These conditions may occur in or adjacent to a stream, creek, or low elevation area.

The mechanisms for development of wetland soils and the reason that wetland plants are found on these soils are related to soil characteristics as influenced by waterlogging.

The chemistry of waterlogged soils changes over time. As explained previously, the oxygen found in water-filled voids between soil particles is quickly exhausted by biological and chemical oxygen demands. Once the oxygen is gone, it is slow to be replaced. Little can diffuse through the water to replace that which is gone, and the anaerobic or "air-free" condition begins.

In the absence of oxygen, the soil electrochemistry changes greatly. No longer is oxygen abundantly available for reduction-oxidation or "redox" reactions (Wetzel, 1975). Depending on the duration of inundation, various elements including sulfur and nitrogen change their compositional forms. Iron, manganese, and other elements are also involved, and they change their "oxidation states" (Foth, 1990).

51

Figure 4.1   These coastal emergent wetlands experience a 3- to 4-in. change in water level every 2 hr during the growing season. The rocking action of changing wind stresses or seiches across the Great Lakes creates this ebb and flow of lake waters, where tides are essentially absent. Pictured are the Straits of Mackinac, Lake Michigan, in Michigan.

The duration of waterlogging has a great influence on the variety and quantity of anaerobic soil chemical products found in soils. Over time, change in the electrochemical environment of soils results in production of different compounds as compared to those found in aerated soil environments. These include different forms of iron and manganese, as well as different forms of other elements, compounds, or ions. Other examples include methane instead of carbon dioxide, ammonium ions instead of nitrates, and hydrogen sulfides instead of sulfate ions (Mortimer, 1941; Wetzel, 1975). Indeed, conversion of chemicals from one form to another is a major value and function of anaerobic soils and wetlands.

The soils will remain largely without oxygen as long as they remain waterlogged. They return to the oxygen-rich or aerobic condition after the water has drained out of the soil and air again fills the voids between soil particles and cracks in the soil. After the water drains from the soil and oxygen is again available in the system, the soil chemistry changes to that of an aerobic system as a result of the electrochemistry of oxygen (Mortimer, 1941; Good et al., 1978).

What remains in the soil horizon is the record of the general presence of water and period of duration of waterlogging. The record is in the form of chemical compounds or deposits in the soil horizon resulting from anaerobic conditions. The remaining chemical products can be used to judge whether soil waterlogging has occurred at a given site at some time of year in the past.

A common condition found in hydric mineral soils is that of iron oxides deposited after air is restored to the soil system following waterlogging. The waterlogged and anaerobic soil conditions initially cause solubilization of iron compounds normally insoluble, and iron ions become available in the anaerobic, waterlogged system. When the water table drops and water drains from the soil horizon, the iron ions and other compounds are converted to oxides in the presence of "new" oxygen in the air. These iron oxides are insoluble in oxygenated water or in the presence of air and are deposited in the voids and cracks of the soils, creating a series of irregularly distributed, "rusty" colors called iron oxide mottlings.

The presence of mottlings is a good indicator that the soils have been waterlogged at some time in the past. They may have been deposited during the nongrowing season or may have been deposited several years ago during periods of higher precipitation, snow-melt, or other conditions that cause water-logging.

Iron oxide mottlings can also occur at a variety of times during the year. So, they are not conclusive evidence of long-term waterlogging during the recent, growing season period. This is because they may have developed during the non-growing period or developed during a particularly unusual precipitation year.

The quantity of mottlings may be interpreted as being some-what related to the frequency of waterlogging. Mottlings vary greatly in size distribution and in color. They may appear as a fine distribution of small particles or may be large, nonrandomly distributed deposits. Mottles may also vary in their color. Their variability can result from local chemical conditions, frequency and duration of flooding, and other variables.

It is desirable to describe whether mottlings are present in the soils and to characterize their color. These mottlings may be characterized by comparing them to the Munsell Color Chart (Munsell Color, 1990) and recording hue, value, and chroma. Conduct this evaluation for all visible soil layers or horizons. It is also desirable to record the distance from the surface to where they are deposited, as the distance may be indicative of the elevation of the seasonal high water table.

Long duration periods of anaerobic conditions produce materials indicative of severely waterlogged soils. For example, the presence of manganese deposits called "concretions" in soil horizons results from waterlogging of long duration in period.

In frequently flooded or constantly waterlogged soils, there are formations present called "gley" deposits or "gleyed" soils. The presence of loose grey soils and grey deposits in soil voids results from long periods of anaerobic conditions. These grey deposits or "gleys" have distinct, though not outstanding, colors. They may be described with a Munsell Color Chart (Munsell Color, 1990) specific for gleys.

Hydric soils are often characterized by a relative abundance of organic matter. This occurs because decomposition of organic matter proceeds at one-fourth the rate in waterlogged soils as compared to soils that are aerobic or aerated (Wetzel, 1975). As a result, the organic matter builds up, and over hundreds of years an organic-rich soil is developed.

Organic matter has a very low reflectance of light and tends to stain soils and make them very dark in color (Lyon, 1987). This dark staining can be viewed by digging up the top 12 or more inches of the organic soil. This dark color is commonly an indicator of hydric soils, and the condition can be judged by low (/2, /1) chromas on the Munsell Color Charts.

Longer-term waterlogging of soils (Figure 4.2) also results in the presence of reduced forms of sulfur and carbon. These sulfur-based compounds include examples that have well known odors. Highly anaerobic conditions can be sensed by the "rotten egg" smell of sulfur compounds. Reduced carbon can be sensed by the "swamp gas" smell of methane.

Sampling these soils is usually a simple matter of digging or probing the soil. In very wet or submerged soils, special techniques may be necessary (Wetzel and Likens, 1990; Mudroch and MacKnight, 1991). These techniques really are more for research than permitting because soils that are very wet or submerged meet the soil hydric criterion and may not require sampling to establish a record of an obvious condition.

SOIL SURVEYS

County Soil Surveys were developed by the soil scientists of the USDA Soil Conservation Service (SCS) in cooperation with state university and state resource agency soil scientists. They are based on extensive field and office work.

In a preliminary evaluation of a property, the Soil Survey may be very helpful in identifying hydric soil types and potential jurisdictional wetland areas. The level of detail of a Soil Survey is good, but necessarily general in nature. The Soil Surveys include soil type maps that consist of soil boundary delineations on copies of black and white aerial photos (USDA, 1962).

Figure 4.2   Standing water for durations of a week or more create anaerobic soil conditions. This condition can be a major regulator of plant species types found in the flooded area if flooding occurs at some frequency during the growing season year after year. The example is an isolated wetland hardwood forest in the glaciated terrain of central Ohio. Note the adjacent upland area and homes.

This "photo map" product can be easy to use because the aerial photo mapping base records a great deal of information. The "photo maps" are a record of actual conditions rather than a "stylized rendition" of conditions typically supplied by map symbols on maps (Figures 4.3 and 4.4).

Soil Survey products also supply detail on cultural or "planimetric" information. These details can be used to locate a given property on the soil survey maps. Once the property is located, the reader should note the presence of soil types indicated by soil type boundaries and map codes. A list of these types should be made, and the characteristics of each soil type should be evaluated from the descriptive part of the Soil Survey.

Soil types are described in the county Soil Survey by the name of a location where the particular soil type was originally described. For example, a Houghton muck was first described near the town of Houghton, a Kokomo soil near Kokomo, and so forth. The system is further detailed in the Soil Survey Manual (USDA, 1962) and other references (Foth, 1990).

The general soil types are also labeled by agricultural soil scientists using the Universal Soil Classification System. It is hierarchical in levels of categorization and its system of labels. The Soil Survey uses these levels of detail to help describe the characteristics of soils.

Under the name of each soil type, the Soil Survey lists a variety of information which characterizes the soil. Included in the narrative of the Soil Survey and in the tables are soil particle size distributions or soil texture. The general hydrological characteristic descriptions of soils are addressed, as well as the agronomic, forestry, and engineering-related soil characteristics. Tables provide information on depths and types of layers or horizons; soil physical characteristics; typical crop yields for the soils; suggested land uses such as farming, forestry, or wetland habitat for each soil; suggested tree or shrub plantings; soil and nutrient characteristics; and a great deal more.

While the information in a county Soil Survey is general, it is of good quality and available for almost all counties in the

U.S. The Soil Survey is uniform in content and detail, and it is free. It is an ideal, routine or reconnaissance level tool for planning and management. Hence, it is often used by groups for that purpose, and it is certainly used as an information source by the U.S. Army Corps of Engineers (USACE).

A disadvantage of Soil Surveys and their level of detail is that many of these products were completed some time ago. Often, users do not recognize this fact and make decisions based on old information. It is particularly difficult to locate a given parcel of property on an older soil survey. Where recent urbanization has occurred, the new streets and new land-marks are not available for reference on the older soil survey maps. Any errors of this nature become obvious with field visits.

Soil Surveys were not developed and produced to function as a regulatory tool. However, they serve the need for general information and enjoy widespread availability.

As a practical matter, the presence of hydric soil types on the Soil Survey of a given property can be a "trigger" for an increased level of attention by regulatory agencies. They do not within themselves present the best product for jurisdictional evaluations and as such should be used for routine level evaluations.

MUNSELL COLOR CHARTS

The Munsell Color Charts (Munsell Color, 1990) provide a good, qualitative tool to estimate color characteristics of soils. The results of physical and chemical processes that create soil types and soil horizons can be recorded in the field using Munsell Color Charts. The color characteristics of soils are conceptualized as color hue, value, and chroma. Hue refers to the general color; value refers to the depth or intensity of the color, and chroma refers to the strength or darkness (Foth, 1990).

A Munsell Color Chart evaluation of soils involves three steps. They include selecting the correct color and hue page in the chart, using the page in the field to identify soil color value characteristic, and identifying the chroma or strength of color along the right axis of the page.

Munsell Color Charts should be used to record the general color characteristics of the soil horizons encountered. One measures the background color of the layer or soil "matrix" color. One might record two such estimates for each sample site; one for the top layer of soil or "A" horizon, and one for the subsoil or "B" horizon.

As discussed previously, certain chemical products of aerobic and anaerobic conditions are manifested as colors displayed by mottled or gleyed soils. If mottlings or gley soils are present, additional records of color should be made to document either of these conditions and to do so for each horizon.

One can find the general colors of a given soil type listed in the Soil Survey. The color swatch can be examined in the book of Munsell Color Charts, and the hue page can be selected. On a given Munsell Color Chart hue page (e.g., 10YR or 5YR), the color choices will be presented as a matrix of values and chromas for each hue.

Use the appropriate hue card to make your estimates in the field. If difficulty is experienced identifying the best hue card to use, one should again check the Soil Survey for the appropriate choice of hue page for the soil types mapped on the property.

If mottling or gley areas appear in the soil matrix, Munsell Color Charts should be used to estimate their color. There exists a special Munsell Color Chart for gley soils alone, which is optimized for the blue-grey colors of gleys. This tool is currently of limited use, as the presence of gleys is indicative of hydric conditions and detailed recording of their characteristics may be of limited interest only. Adequate detail can be supplied with the soils Munsell Color Charts, if the special gley chart is unavailable.

The characteristic colors of a given wetland soil and its soil layers or horizons are recorded in the county Soil Survey. In general, soil color chromas of /1, /2 are indicative of wet soils and a hydric condition. The Federal Manual and other manuals indicate soils with these chromas should be categorized as hydric, and they meet the soil hydric criterion for jurisdictional wetland areas.

PLANT MEASUREMENTS

A given plant's abundance and distribution will be largely determined by its tolerance to wetland soil and hydrological factors. As a result, wetland plants are a valuable indicator of wetland conditions in a given area. Plants respond to many environmental parameters such as flooding, topography, soil type, soil nutrients, and general water quantity and quality conditions. Because of this capability to grow under these environmental conditions and variability of habitats they grow in, plants can be used as indicators of potential wetland areas.

Often the change in wetland plant types across a wetland area is a gradual one, with overlapping areas of different plant species that tolerate slightly different wetland soil and hydrological conditions. The presence or absence of certain plants can be used to help classify these wetlands as to type and to map these types of wetlands.

For regulatory purposes, the dominant plants need to be identified at the species level (Figure 4.5). Dominance is defined in the Federal Manual and other manuals as the relative abundance of the most common plant species in a given layer of vegetation. The abundance of each dominant plant species needs to be estimated in the field.

Only the most dominant species need to be estimated. However, it is good to note the relative abundance of all plants in the sampling area. This would be true of any obligatory (OBL) or facultative wetland (FACW+) (Reed, 1988) plants, which may not be dominant but are certainly of interest.

The tree, shrub, and ground layers of vegetation should be sampled (Figure 4.6) if plants are found in each of these layers. One method is to describe the type and to estimate the relative quantity of plant species and their dominance. Dominance can be estimated by visual determinations of percent areal cover of the plant layer or stratum, or number of stems/trunks in a given vegetation layer.

Relative abundance can be estimated in 5 or 10% increments totaling a 100% relative cover or abundance for a given sampling location. This can be done by visual estimate, by counting stems/trunks, or by relative size or girth of the tree species. The level of detail (5 or 10% increments) of vegetation can be selected by the user.

Figure 4.5 Note the dominant plant species in the tree layer is Tamarack or Eastern Larch (*Larix laricina*), a common species found on the wetland edge of ponds in the lake states and other places.

Figure 4.6 This desert river displays the three major layers or growth forms of plants found in a vegetation community: the ground, shrub, and tree layers. It also shows how one or two of these layers may be missing from certain parts of the area. The location is the San Pedro River west of Tombstone, Arizona.

The Federal Manual, other manuals, and experience of the author suggests that estimates for the tree layer should be made within 30 ft of the sample point. A 20-ft radius circle from the sample point should be sampled for the shrub layer, and 10 ft in each direction from the center of the sampling point for the ground of herbaceous layer.

Estimation of relative plant dominance is potentially difficult. The delineator needs to develop a suitable approach, and one that is acceptable to the USACE. The approach must also be repeatable, and should be performed in a direct and timely fashion.

Knowledge of plants is highly desirable, but one need not be an extensively trained botanist. The Federal Manual (FICWD, 1989) states that the person evaluating the plants should be able to identify the dominant plant species present in the area (Figure 4.7). A general level of familiarity with plants is important, and this knowledge can be supplemented by plant books.

A variety of plant identification books should be used for identifying or "keying" plants by their species names. One should make use of local plant books, as they often have maps of plant species distributions. Books with color photographs of the plant species of interest can be invaluable, along with plant identification keys.

In the beginning of any effort, it may be desirable to collect plants in the field. A good approach is to refrigerate them until there is time to identify the plants at the office. Once an individual is thoroughly familiar with the dominant plant types, they may be described in the field without collecting plants at each, individual sample point.

One needs to employ photos in addition to traditional "keys" to help ensure a correct determination. The combination of local plant distribution lists or maps, plant photos, and plant keys will greatly assist the reader in correctly identifying plant species.

Due to the inherent difficulty of plant identification for nonbotanists, it is important to use a number of books and sources. At a minimum, one needs a wetland plant book (e.g., Fassett, 1957; Hotchkiss, 1972), a tree book (Harlow, 1957; Sargent, 1965; Miller and Lamb, 1985), a shrub book (Billington, 1968), books about nonwoody plants (Dana, 1963; Peterson

Figure 4.7 Plant identification is difficult because wetlands have a combination of upland, wetland, and aquatic plants. This can be a greater challenge where wetlands and wetland plants are ephemeral. Try to make plant and wetland determinations during the season when the indicators can be expected to be present. Here is an ephemeral stream that may be a wetland in Valley of Fire State Park, Nevada.

and McKenny, 1968; Courtenay and Zimmerman, 1972; Mohlenbrock, 1987; Forey, 1990), books on grasses and sedges (Hitchcock, 1971; Knobel, 1977), as well as books about a variety of plants (Britton and Brown, 1970; ARS, 1971).

A variety of books are necessary to address the mix of upland and wetland plants that occupy the interphase between each environment. Delineation of wetlands occurs at the edge, and hence a mix of publications is required. Additional books that can be helpful in plant identification are listed in the Bibliography.

Plants typically have one or more names in common usage, known as common names. These may be used in describing the plants in the field and/or in the report of the project. Plants also have a scientific name. It is best to employ scientific names in the wetland report and to do so on the record sheets (e.g., Appendix B) supplied in the Federal Manual (FICWD, 1989).

The scientific names have two parts written in Latin. The first word is the family or generic name. Plants are grouped into families thought to have a common evolutionary ancestry. This is based on plant physical characteristics or morphology. The second name is that of the individual plant or plant species in the family.

Until familiarity with common or scientific plant names is established, use general names or identification labels in the field. This labeling approach allows identification in the field and allows one to record the dominance level, and then determine the actual species scientific name later in the office.

Consider drying selected examples of plants. This may be done in a plant press of layered cardboard and blotter paper, all secured by straps to flatten and help dry the plants. These plants may be useful at a later time, as a record to verify their identity and satisfy regulators.

The plant criterion for a jurisdictional wetland is determined by enumerating only the dominant plants that have the highest probability of existing in wetland areas. These are the FAC, FACW, and OBL categories of plants. A given area is determined to be a wetland for this individual, jurisdictional criterion when the total abundance of FAC, FACW, and OBL plants exceeds 50% of the total frequency of dominant plants found at a given sampling location on the site.

A fundamental problem for analyses of wetlands is that some plants "evade" harsh winter conditions through death or dormancy. As a result, there is an absence of structures like leaves or flowers to assist in their identification. The absence of these and other plant features can make plant identification difficult. Plant species keys have been developed for winter conditions (Harlow, 1946; Symonds, 1958; Trelease, 1967; Symonds, 1973) and can be employed, but identification is much slower. It may be impossible to identify or locate some plants, particularly on the ground or herbaceous layer, if the plants have decomposed or are otherwise absent.

5

Additional Methods and Considerations

Wetlands exhibit a variety of conditions that may not be easy to identify and characterize. Conditions may require application of methods that supply detail beyond that presented in previous chapters.

The following methods and considerations may be very helpful in a variety of situations. Some of these methods may be comprehensive in detail and their implementation, along with intermediate level examples of methods, will facilitate the assessment of wetlands and their variable characteristics.

The identification and delineation of jurisdictional wetlands should employ a combination of methods. Here, information is supplied on individual methods and scientific or engineering approaches that may be useful in a variety of situations. The best approach for evaluation of a given property would be to combine routine and intermediate level techniques and to draw upon the following methods to supply additional information as dictated by the problem at hand.

TOPOGRAPHIC MAPS

Large scale topographic maps are usually available for sites to be developed. These large-scale, engineering-style maps are very useful for delineation of wetlands. They present topographic detail of much higher frequency and quality than U.S. Geological Survey (USGS) 1:24,000 scale maps.

Use of the topographic maps allows the relative topography of the site to be evaluated. This fosters identification of low spots where water may pond and assists in the identification of streams, creeks, gullies, flood plains, and other conveyances of water. These water-formed landscapes or low areas may be potential wetland areas, and it is desirable to use maps to locate them and investigate their presence in the field.

Large scale topographic maps are commonly made by surveying and photogrammetry techniques. Generally, they are produced at scales of 1 in. = 50 ft, 1 in. = 100 ft, or 1 in. = 200 ft. They are called "large scale" because the ratio fraction of scale is large, as compared to the ratio fraction of a 1:24,000 scale map (1 in. = 2000 ft). A method to remember these terms is that objects appear "larger" on large scale maps.

The ranges in vertical elevation of the landscape are presented as contour intervals. They can be used to characterize topography from these large scale maps. One can also evaluate topographic detail from 1:24,000 scale USGS quadrangles, but large scale topographic maps present more detail.

Topography is commonly presented either in 1- or 2-ft contours. In mountainous regions or for maps of small scale, the contour interval may be 5 or 10 ft. The actual interval depends on the general topography of the site and the needs of the project.

Contours of 2 ft are fine for general design of projects. A 1-ft contour is necessary for finer details of project design including cut and fill calculations, planning and design of storm sewers and water lines, and the platting of property for homes. Either contour interval on the large scale topographic map can be employed in wetland delineations.

Large scale topographic maps can be valuable for presentation of the location of jurisdictional wetlands. This product makes an excellent record for permitting when maps of 1 in. to 100 ft or 1 in. to 200 ft scale or larger are used. Such maps and wetland delineations may also facilitate highly detailed determinations of wetland areas.

Large scale maps also indicate where detailed "intermediate level" sampling was conducted and facilitate later inspection of the site by regulatory personnel. They can also be employed to record the extent of hydric soils as determined in the field. In general, they make a very good "base map" and can be annotated with appropriate detail to make a product that is very useful in permitting.

The USGS topographic maps are available for the U.S. at a variety of scales. The largest example is the 1:24,000 scale quadrangle, and these maps can be found for most of the continental U.S. The USGS quadrangles may be used in reports to identify the general location of the site; the presence of adjacent streams, ponds, and rivers; and general location of potential wetlands. They are also good for identifying adjacent resources of note, such as other large or significant wetlands, or presence of navigable waterways, or other large or significant waterways such as rivers and streams.

Many people make use of these 1:24,000 scale maps in wetland reports, as the maps are nice products and readily available. The resolution is really too coarse, however, for recording the size of wetlands encountered in permitting actions. They are inferior to large scale maps for making determinations of wetland size or acreage. Potentially, large errors can be made due to the course resolution associated with the scale. They are certainly useful in the absence of aforementioned large scale, engineering-style topographic maps, but should not be substituted for such large scale maps.

It is also desirable to obtain general purpose road maps of the region and use them to locate the property very clearly for the reader. This is accomplished by supplying a copy of the road map with the property location indicated. Road or county maps are also useful in developing a verbal description of the property location, i.e., a description of the property as it is referenced to streets. Certainly, a legal description of the property is also valuable, though such a detailed description may not be necessary for some activities including permitting.

Other types of maps, or reports containing maps, may be encountered in the search for available information. As stated previously, it is important to make use of all available data sources in making an assessment. Hence, one should use all mapping information to ensure a complete data base for analysis. This approach also allows one to take advantage of available and "free" information and ensures that no important mapping or wetland details are missed or ignored.

AERIAL PHOTOS AND REMOTE SENSOR DATA

In a similar manner to maps, aerial photos provide a vertical or synoptic view of wetlands (Roller, 1977; Lyon, 1979; Lee, 1991). Depending on the altitude where the photographs were taken, photos can be used to identify wetland plant species and/or groups of plants known as wetland plant communities (Figure 5.1). Photos can also be interpreted for hydrological conditions (Lyon and Drobney, 1984; Lyon et al., 1986a; Lyon and Greene, 1992) and provide a valuable record and source of data for assessment of the year-to-year or season-to-season variability in wetland hydric conditions (FICWD, 1989).

Many wetland types exhibit distinct light reflectance characteristics in the visible or infrared portions of the electromagnetic radiation spectrum. Wetland soils and water have distinctive reflectance characteristics that can be used to identify their presence and condition. Distinct shapes or patterns are also found in association with wetland site conditions. All these characteristics can be used for identification, and they are evident on aerial photos as different colors, tones, shapes, and textures or patterns (Lyon, 1979; 1981).

Aerial photos are also valuable as a historical record. They are a permanent measure of conditions at one point in time. Aerial photos have been acquired of the U.S. on a regular basis since 1935. It is common to find a series of dates of photographic coverage for a given site, beginning before World War II and continuing on a periodic basis to the present (Lyon, 1987). Often more than ten, individual dates of aerial photos can be identified for a given site.

This photo record can be exploited whenever the historical condition of a site needs to be evaluated (Lyon, 1987). They are also very useful in judging change in conditions over time (Lyon et al., 1986a; Lyon et al., 1986b) or for making an assessment of the original conditions of wetlands.

From a regulatory viewpoint, the original condition and subsequent illegal filling of a wetland can be documented from photos. Most U.S. Army Corps of Engineers (USACE) commands make frequent use of historical and current photographs of wetland areas. Filled areas look very different from

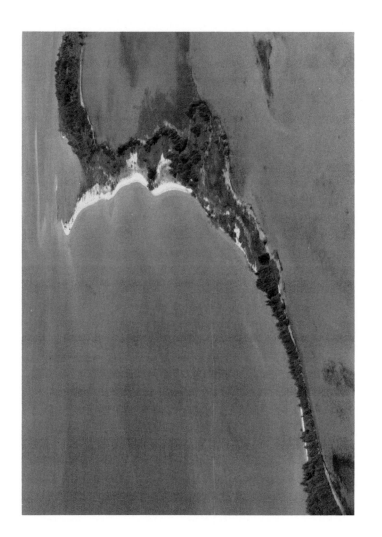

Figure 5.1 A low altitude aerial photograph captures the variety of coastal wetlands surrounding this island in Lake Michigan. Submergent vegetation is evident as dark-toned areas beneath the water surface. Emergent wetlands have a uniform dark tone and are found on the shore areas. Forested wetlands are very dark toned and of rough texture from the tree canopies.

undisturbed wetlands, due to the difference in reflectance of dry soil and wetland. Simple comparisons of current and historical photographs allow rapid identification of filled areas. Subsequent evaluation of permit files and/or field inspection, can verify the character of the fill.

An additional advantage that aerial photos offer is the capability to make precise and accurate measurements using principles of surveying and photogrammetry. From a few survey measurements on the ground, or distance measurements from maps and measurements from aerial photos of the same distance, one can calculate horizontal distances and vertical elevations of selected features. Almost all topographic maps are made with these photogrammetric and surveying technologies, and hence, one can develop topographic mapping products from photos and use the variable of elevation to help in delineating wetlands.

The identification of potential wetlands at the routine or intermediate level can be difficult at certain times of the year. This is due to inclement weather including snow fall, cold and hot temperatures, or some inherent problems such as size of the properties, presence of hazardous materials, limits to access of lands, and/or possible disturbance of plants or animals. Other considerations are the risk to field personnel from noxious plants and potentially dangerous wildlife or insects. Photos can be used under these conditions allowing "access" to the site. They may also augment other measurements and allow some work to progress under less than desirable conditions.

One can use aerial photos to evaluate wetland characteristics over time (Lyon and Olson, 1983; Lyon and Drobney, 1984; Lyon et al., 1986a; Lyon, 1989; Williams and Lyon, 1991; Lyon and Greene, 1992) and do so during the course of the senescent or dormant seasons. Many groups make use of aerial photographs to identify and quantify potential wetlands year-round and avoid poor conditions for field work (Figure 5.2). For the USACE permitting purposes, it may still be necessary to await the growing season to complete the analyses and present the results for inspection.

There are numerous advantages in employing analyses or interpretations of photos as one of the many data sources in an

Figure 5.2   Forested wetlands can be identified from this low altitude aerial photograph. Note the shape of the tree canopy or its shadow can help identify the trees as to species. Emergent wetlands are also evident from their dark tone and the coarse texture the plant stalks make as they break the water surface.

integrated wetland analysis. In the process of identifying wetlands, it is desirable to determine what types of wetlands are
present. Using aerial photos, it is possible to separate the
wetlands as to type such as forested wetlands (Figure 5.3),
marshes, and/or riparian areas. Using aerial photographs,
wetlands can be labeled or categorized as to types according to
a given classification or categorization scheme. Classification
schemes or systems are useful in many activities including
delineation. In particular, scientists and engineers favor the
USGS Anderson system (Anderson et al., 1976) or the National
Wetlands Inventory System (Cowardin et al., 1979) for describing wetlands as to type.

Often individual states have optimized the Anderson system for local conditions, and this state-driven categorization
of wetlands usually has a similar goal and methodological
approach to that used in the U.S. Fish and Wildlife Service
National Wetland Inventory (NWI) program and other state
or federal agency programs.

In these efforts, wetlands are identified by the interpreter
from aerial photographs, and the wetland boundaries are plot-
ted. The type of wetland is then described according to the
types listed in the applicable classification system such as the
state developed system. Often they are also described with the
NWI classification system to facilitate comparisons and communications of results (Cowardin et al., 1979).

One can acquire aerial photos by custom order or contract
through a vendor. Or one may make use of a variety of dates
and types of aerial photographs from governmental and private archives (Lyon, 1987).

Archival photos will be of both historical and of recent
origin. It is desirable to obtain and use a variety of photos from
the archive (Appendix A). This is because one can accumulate
historical data on wetlands and other conditions related to
soil, hydrology, and vegetation conditions. One can also document the chronology of land use activities and do so with an
independent source of information.

Aerial photographs are acquired at different times and record
a number of wetland conditions. These include hydrological
conditions, as well as the extent and type of wetlands present
(Figure 5.4). Use of classification schemes and aerial photo-

Figure 5.3 From higher altitude, the expanse of wetlands can be visualized. Wetland types include submergent, emergent, shrub/scrub, and forested wetlands in Wilderness State Park, Straits of Mackinac, Michigan.

Figure 5.4 Low altitude photographs may be taken with a 35mm camera and a rented aircraft, such as this photo. Note the detail in the tree canopy, which can aid in tree species identification, and the small emergent wetlands of dark tone found in the light toned, marl bottom lake in Wilderness State Park, Michigan.

graphic interpretation of wetlands provides a means to map general types of wetlands. The resulting product can be used to determine the area of the wetland.

It is important to consider the time of year when the particular archival photos were taken and the prevailing conditions of plants, soils, and hydrology at that particular time. If one accesses a variety of archival photographs, it is also possible to evaluate wetland conditions during "leaves-off" and "leaves-on" periods of the year. Thus, seasonal change can be inferred from multiple date coverage of wetland areas.

One needs to interpret the aerial photographs with knowledge of the seasonal condition of plants. This knowledge of condition is often referred to as the "growth cycle," "seasonality," "phenology," or the "crop calendar" of the plants.

There are certain common elements of plant "behavior" or growth cycle such as death or senescence. Senescence or "fall colors" occur when the chlorophyll and its green reflectance characteristics disappear with the breakdown of the chlorophyll. We see the reflectance of other plant pigments that remain. These other pigments include anthocyanins, flavonoids, and other constituents. These pigments are plant wastes stored in vacuoles or waste-holding voids within the plant cells. These compounds give deciduous foliage its yellow, orange, and red colors in the fall season or provide the orange color of dead evergreen needles.

Senescence is a common characteristic of seasonal plants, and it is a common element in the behavior of all plants. One can also observe genus and species differences in the growth cycle including differences in the timing of reproduction or flowering, plant shape or structure, rates of growth, timing of the on-set of senescence, and differences in location where a plant grows or its habitat (Figure 5.5). Many of these plant characteristics can be interpreted and recorded from aerial photographs.

Sources of these photos are described in the Appendix B. They include the U.S. Geological Survey (USGS), the U.S. Department of Agriculture (USDA) and related agencies, the National Oceanic and Atmospheric Administration, the U.S. Environmental Protection Agency (USEPA), the U.S. National

Figure 5.5 The lighter-toned area is nonwoody vegetation found on point-bars and in abandoned river channels or oxbows. The dark-toned and coarse-textured areas are coniferous forests composed of White and Black Spruce, and they may be wetland areas. This scene is from interior Alaska just south of the central Brooks Range.

Archives and Records Service (NARS), as well as the archives of mapping companies (Lyon, 1987; Lyon and Greene, 1992). These groups maintain coverage over portions of the U.S. and can be counted on to supply multiple dates of coverage of a given area.

Photos are also available from local archives, and these may be maintained by a variety of groups. In each county seat of government, aerial photos of recent "vintage" are held by the county offices of USDA-Soil Conservation Service (SCS). These are the same offices that supply the county Soil Surveys. In the county offices, the photos may be viewed, and assistance can be provided in ordering them from archives (Figure 5.6).

In the state capitol and in regional headquarters, photos may also be available. In particular, one should try to access archives maintained by state Departments of Transportation (DOT) and Departments of Natural Resources (DNR) or state agencies with similar mandates. In particular, state DOT photographs are valuable because they are flown at low altitude and they are large in scale (Figure 5.7). These products supply very good detail for local wetland areas.

Contract photography missions to acquire "custom" photos can be flown by a variety of groups (Figure 5.8). Commercial mapping firms can be employed, and these regional and national firms can be located in telephone directories. Many of these firms can also be identified from the list of sustaining members of the American Society for Photogrammetry and Remote Sensings (ASPRS). This list is available from the society (5410 Grosvenor Lane, Bethesda, MD 20814) or from its journal, "Photogrammetric Engineering and Remote Sensing." This journal is archived in most libraries or is available from the ASPRS.

As described earlier, most large scale topographic maps are compiled from aerial photos. These photos can be obtained from the company charged with mapping the site for engineering design purposes. The scale of photos should be large, and hence, they will provide additional detail.

One can also take advantage of the capabilities of color and color infrared types of films. Often these films reveal more information about plant and water resources than do black and white photos alone. Portions of the spectrum that humans

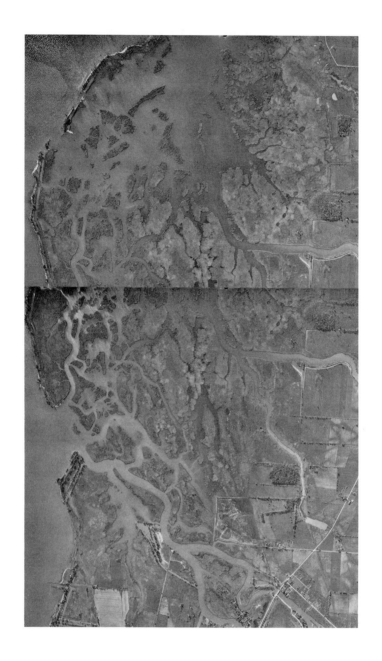

Figure 5.6 These approximately 1:20,000 scale photographs are of western Lake Erie Pointe Mouillee wetlands near Detroit, Michigan. The photos were taken by the U.S. government and represent the type of images available from archive (Appendix B).

Figure 5.7 These approximately 1:8000 scale photographs are of an island in the mouth of Sandusky Bay, Lake Erie, Ohio. Note the detail apparent from low altitude, engineering style photographs taken by the state Department of Transportation. Note the presence of coastal beach wetlands (US) and lacustrine emergent wetlands (EM). Inland, there are emergent wetlands, shrub/scrub wetlands (SS), as well as upland forests. These photographs are arranged for stereoscopic viewing and interpretation of relative elevational differences in support of wetland assessments.

Figure 5.8 These vertical, large scale aerial photographs were taken with a 35mm camera. Note the detail available from low altitude coverage and the relative elevation differences evident from stereoscopic viewing of overlapping aerial photographs. Present are beach wetlands (US), aquatic beds (AB), emergent wetlands (EM), and cedar swamp vegetation (FO). The location is Wilderness State Park, Straits of Mackinac, Michigan.

may not see, such as the infrared, can be recorded with special film emulsions or instruments. Color or color infrared photos are available from archive, by contract with an aerial mapping firm, or photos may be taken by other means (Lyon, 1987).

Several characteristics of wetlands can be used to identify and quantify their presence from aerial photos. Color infrared (CIR) photos are particularly suitable for identification of wetland areas, as they can help distinguish plant species or plant communities depending on scale.

One can also identify water resource characteristics using CIR photos. A valuable application is locating the edge of the water surface. This can be accomplished by taking advantage of the very high absorption characteristic of infrared light by water and the contrasting very high reflectance of soil and vegetation. Reflectance of light is defined as the ratio of the incoming radiation to the exiting radiation, and it is characteristic of a given material or mixture of materials. Reflectance is also known as the tone we see on aerial photographs, and reflectance is probably the most important clue in identifying features on the earth's surface.

Experience has demonstrated that several wetland plant communities or types can be identified from CIR photos using photo interpretation "clues." These types include emergent wetlands, submerged or submergent wetlands, shrub/scrub wetlands, and forested wetlands. The clues used to identify these wetland communities are described as follows.

Emergent Wetlands

Emergent wetland communities are variable in plant species composition and as a result will be variable in color or tone on infrared photos. This variability in color or tone is distinct from the uniform color associated with other plant communities such as stands of trees or farmed fields composed of one type of plant.

Sometimes, one encounters uniform tone or color when a single species is predominately present in a vegetation community. On the eastern marine coast, *Spartina* are found in "monotypic" stands, and these wetlands are usually uniform in color or tone. In areas of cattail stands, one will encounter somewhat uniform color or tone, but these clues may be broken up by the clonal growth pattern or "bunch-like" groupings of cattails. The "bunches" are created by vegetative reproduction from existing plant material, as opposed to the seed or sexual methods of reproduction.

Emergent wetlands can be identified by a number of photo interpretive clues. For example, their proximity to the terrestrial/aquatic interphase can be a clue to their presence. Also the presence of streams entering and/or exiting the site is a good clue.

The red or magenta color on CIR photos is indicative of growing plants. One can note their irregular boundaries, in contrast to human created boundaries such as the regular pattern of property lines that follow the U.S. Land Survey.

During the nongrowing season, emergent wetland areas appear very dark in color or tone. This is because the plants are devoid of chlorophyll, and hence, the characteristic reflectance of green plants as shown by a red color on CIR film is absent. The presence of water also makes the reflectance of emergents or their residue even lower than surrounding terrestrial plant and soil materials.

The combination of chlorophyll-free plant residue and presence of water creates this very low reflectance or dark color/tone condition in dormant emergent wetlands. This characteristic reflectance can be used to reliably identify wetland areas on aerial photos.

Submerged Wetlands

Submerged wetland communities are found beneath the surface of the water and hence are called "submerged" plants or submergents. They appear characteristically dark in color or tone (Figure 5.9). This is due to the inherent low reflectance of water that covers the plants. The dark color or tone is also due to the general low reflectance of plants in the visible portion of the spectrum. Most infrared light is extinguished by water, and hence, the infrared reflectance of plants beneath the surface of water is very low to zero. There is no infrared contribution or red color found on CIR photos of submerged wetlands.

Figure 5.9 Submergent wetlands are found throughout the image, particularly at the top one third of the image, and are detected by their dark tone and shape in Wilderness State Park, Michigan. At the top center, three people are walking back from fishing for Small Mouth Bass. The bottom third of the image is the same location pictured in Figure 5.8.

These phenomena create the very low reflectance, color, or tone of submergent wetlands. They are some of the darkest features on aerial photos, and they can be identified by the characteristic dark color or tone and irregular shape. This is in comparison to sand and clay bottom sediments, which have higher relative reflectances (Lyon, 1980; Lyon and Olson, 1983; Lyon et al., 1992). They are commonly found in shallow water where the high reflectance or bright tone of bottom sediments is apparent from the air and provides a stark contrast between the dark-toned plants and light-toned sediments and irregular shape of plant community boundaries.

Shrub/Scrub Wetlands

Shrub/scrub wetland areas exhibit wetland soils and hydrology, but differ from the other wetland types mentioned here in that they are populated largely by shrubs, bushes, or brush. These woody plants are persistent year to year. These areas can be identified by the interpretation of their characteristics on photos.

Clues include the rough texture or fine-scale variability in the photo color or tone resulting from the shape of the shrub branching pattern and the shape of the tops or crowns of shrubs. The irregular shape of the wetland area (Figure 5.10) and the presence of water or streams running into and out of irregular-shaped areas are also good clues. The association of the shrub wetland with adjacent emergent or forested wetlands is an additional, valuable clue.

Forested Wetlands

Forested wetlands are commonly know as swamps and are defined by the presence of trees, which mostly cover the site and are persistent. Forested wetlands may be identified on color-infrared films by the red or magenta color of green leaf material in the growing season and by the texture or "roughness" of the tone or color resulting from the spatial variability of tree canopies or crowns. On black and white photographs, forests exhibit the same textural characteristics, but are dark

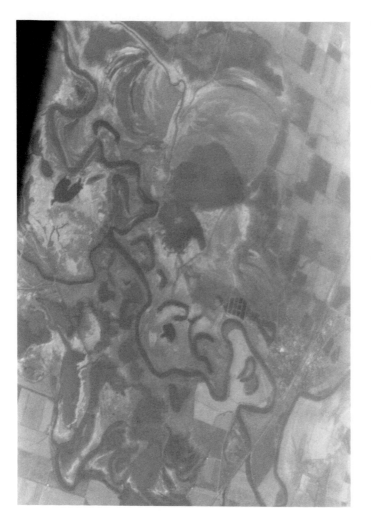

Figure 5.10 This photo was taken from a commercial jet aircraft flying over Wyoming. Note the large size of area where river meandering has created a variety of soil and water conditions such as oxbow lakes. The extreme variability in shapes and the lack of human-induced boundaries such as roads is strong evidence that one is viewing a hydrologically-shaped landscape. The drainage pattern is very disorganized due to the river meandering.

toned due to the low reflectance of woody plants in the visible portion of the spectrum (Lyon and Olson, 1983).

Other good photo interpretation clues include the irregular shaped boundaries of the forested area, the presence of standing water on the ground surface, or a stream entering or leaving the area. The lower relative topography as compared to the general topography of the surrounding areas is an additional, valuable clue.

Forested wetland areas may be populated with trees that drop their foliage in winter (deciduous) or trees with year-round green foliage (evergreens). Evergreen trees will display the characteristic red or magenta color on CIR film on a year-round basis. Hence, CIR photos can help to identify evergreen-forested wetlands and separate them from deciduous-forested wetlands using the contrasting colors or tones of photos from the summer or "leaves-on" period and winter or "leaves-off" period.

Forested wetlands may be difficult to identify with "leaves-on" photographs. Most CIR photos are taken during the growing season, and often the forest canopy obscures the wet ground below. At the height of the growing season, upland trees and wetland trees may be difficult to distinguish based on plant reflectance alone.

Hence, it is desirable to obtain additional "leaves-off" archival photography to help identify forested wetlands from upland forests. Most existing archival black and white photographs were acquired during "leaves-off" conditions, so that ground contours could be mapped without being obscured by foliage. Use of a combination of photos from different times is always a useful approach, but it is particularly so in the case of separating forested wetlands from upland forests.

AERIAL PHOTOS FOR CHARACTERIZING SOILS

Aerial photos and remote sensor data can be useful for identification of wetland soil areas. It is possible to determine general moisture conditions of soils by color or grey-tone of photos. Differences in color or tone may be reliably interpreted for variation in the relative hydrological and textural characteristics of soils (Lyon, 1987).

Wetter areas of soils appear as black or dark tones on black and white aerial photographs. Wet soils are always "darker" than the same dry soil. This is because water greatly reduces the reflectance of materials it coats or mixes with, and brings its low reflectance characteristics to the mixture. This phenomenon can be used as an indicator of relative soil moisture from aerial photographs.

These dark tones can also be used as indicators of relatively low topographical areas, which may commonly be filled with surface or subsurface water and function hydrologically as conduits of runoff (Lyon, 1987). Fine textured soils composed mostly of silts and clays can also be identified by evaluation of photo tone. These soils are usually medium grey to dark grey in tone. These soils may be found on lower elevational parts of fields or other areas where water ponds.

Coarse-textured soils appear as lighter grey tones on black and white aerial photographs. This results from the relatively rapid drainage of rainfall by the coarse-textured soils. Identification of lighter tones in soils may indicate the location of coarse-textured soil deposits in abandoned stream channels adjacent to existing streams and rivers (Lyon, 1987). These abandoned channels may be possible paths of movement of subsurface flow.

LARGE AREA WETLAND EVALUATIONS

Aerial photos are particularly useful in evaluations of potential wetland resources over large areas. An effort might involve areas in size from 2000 acres to an entire state. Naturally, this sort of activity would be of interest to a variety of users. It can assist the regulatory process in general by providing additional information of a uniform quality and content. A similar process is performed by the National Wetland Inventory (NWI) of the U.S. Fish and Wildlife Service to provide NWI maps of selected regions (Cowardin et al., 1979), and these products will help meet the need for generalized wetland information over large study areas.

It is important to note that one should attempt to identify only potential jurisdictional wetlands using a large area evaluation. This sort of a project may be directed to identify areas that are jurisdictional wetlands, but the approach is much more successful when one attempts to evaluate what is called "potential jurisdictional wetlands." To make a true determination of a jurisdictional wetland, it is necessary to test for the three criterion and to do so with field work.

To identify resources in a large area, it is best to use a combination of maps, "custom order" aerial photos, and archival aerial photos. The photographs satisfy many needs, as they will be both of historic and recent origin. It is desirable to both fly "custom" photos and to use a variety of available photos from archive because one can accumulate historical data on wetlands during different times, during different meteorological and hydrological conditions, and during "leaves-off" and "leaves-on" conditions.

One can also take advantage of the fact that archival aerial photographs often include different film types that may reveal more information than black and white photos alone. These color or color infrared archival photos can be acquired for little cost, yet they can potentially supply a very useful "data point" on plants, soils, and hydrology.

The steps in aerial photo analysis of wetlands over a large area would include the following steps:

1. Fly aerial photo coverage of the region or county at relative large scale in fall and/or spring seasons. This will allow evaluations during two different soil moisture and plant growth conditions. Record current land cover or use of the area. Black and white film can be used during "leaves-off" conditions. Color infrared film can be used during "leaves-on" conditions, and one can exploit the special detail CIR photos provide for evaluation of plant species and hydrological conditions.

2. Obtain available, archival aerial photographs from a number of sources. To obtain the most complete record and to ensure aerial photo coverage that addresses all three wetland criterion, it is important to write, phone, or visit the local USDA-SCS offices, state resource agency offices, state transportation agency offices, the USGS EROS Data Center archive, the USDA-ASCS-Salt Lake City archive, and the National Archives and Record Service. Included in the search would be archives of aerial mapping firms and any

other archives that come to your attention during the search. Appendix A presents additional information such as details on agency capabilities and mailing addresses.

3. Take all the photographs, the county Soil Survey, topographic maps, and any other pertinent data sources, and study them to identify and delineate potential wetlands.

4. If custom aerial photos are flown during the growing season, these photos and additional photos from archive should be used to further delineate individual wetland areas.

5. Use the information developed from evaluations and interpretations, along with any field experience, to integrate these data sources. Sources such as maps of geology, watersheds, and drainage can help identify potential areas of wetlands in a large area inventory.

6. Use field work to check the adequacy and accuracy of mapped wetland areas. Make corrections as needed, such that the product shows the highest probability areas of potential jurisdictional wetlands.

   If possible, develop a method to assess the accuracy of these mapping products and the wetland classes they display. This is particularly important if the resulting large area mapping product is to be used by a variety of groups or is to be used as some type of regulatory tool.

7. Take these map products, and transfer boundaries to acetate, mylar, or other permanent, transparent materials to facilitate production of final map products.

8. Provide maps, other data products, and additional details in a report summarizing the methods of production. It is also desirable to detail the characteristics of the products, and estimate their accuracy, if possible.

9. Allow a period for evaluation of the products by interested parties.

10. Incorporate comments, criticisms, and any new information that has been identified. Produce the final products that identify and delineate potential jurisdictional wetland areas.

It is also possible to utilize satellite-based sensors to inventory potential wetlands (Lyon, 1978; Lyon, 1979; Lyon et al., 1988; Lyon, 1989; Lyon et al., 1992). This satellite-based sensor approach is particularly appropriate when large areas are to be evaluated. Most of these satellite sensor data are in a computer compatible format, and the capabilities of computers can be

in the U.S. and hence are of limited value if they are unavailable in your area. This will change soon with the completion of this mapping effort.

A graphic displaying the 1:24,000 scale maps of the U.S. that are available can be obtained from the NWI (Appendix A). Their address is National Wetlands Inventory, 9720 Executive Center Drive, Suite 101, Monroe Building, St. Petersburg, FL 33702.

The wetland maps of 1:100,000 scale are of limited value due to their small scale. Wetlands that are large enough to engender regulatory attention may not show up on the 1:100,000 scale products, but will probably appear on 1:24,000 intermediate scale products.

In production of NWI products, the types of wetlands are categorized using a classification system developed especially for the effort. The combination of delineating wetland areas from one date of photos and categorizing them as to type results in a product that is a thematic map of wetlands. Geology maps, Soil Survey maps, land use cover maps are all examples of maps that display "themes" and are thematic maps.

Wetland types or themes are categorized and described using the National Wetland Inventory Classification System (Cowardin et al., 1979). The system is hierarchial in that it possesses tiers or layers of categories or types. It is possible to make wetland maps of varying degrees of detail by providing all the category labels or only a subset of the category labels (Figure 5.11).

The basic level of identification involves five general categories or types of wetland. The class names are similar to the hydrological conditions of the wetlands. The wetland types and their system labels or types include lacustrine or lake-based wetlands (L), riverine or wetlands found along rivers (R), palustrine or wetlands found inland from obvious water bodies (P), estuarine wetland types found where fresh and salt water systems mix in the coastal zone (E), and marine or salt water wetland types found adjacent to or in saline waters (M).

The NWI map products have been generated for over 15 years and are now selectively available for much of the U.S. If a map is available for a given area, it may be useful as a guide to where potential jurisdictional wetlands are located and provides information as to their identity or type.

One should also note that NWI maps at either 1:100,000 or 1:24,000 scale are necessarily a "coarse look" at the issue and may include too many or too few areas that are true jurisdictional wetlands.

The definition of jurisdictional wetlands and the wetland types used for the NWI are not the same. The NWI describes general types and locations and is on par with the detail and quality of a county Soil Survey. Hence, NWI products provide a good indicator of wetlands for the routine level of wetland evaluation. A NWI map should be used much like the USDA-SCS county Soil Survey products, in that they are an indicator of potential jurisdictional wetlands.

The NWI maps were not intended to be a regulatory tool. They certainly may be used by a variety of groups including the USACE to identify properties that may need a wetland evaluation. Like county Soil Survey data, NWI maps will often help "trigger" an inquiry by the USACE upon development of a property.

SURVEYING AND MAPPING

An important component of a determination of jurisdictional wetlands is an estimate of total wetland area and a map of the location of the wetlands. This information and products may be developed by the wetland expert, if the person is also familiar with surveying, mapping, and photogrammetry. Regardless of background, the wetland expert must be capable of rendering an accurate and precise wetland boundary on the ground and recording that boundary on available maps.

An additional advantage of surveying the boundaries of jurisdictional wetlands is that the acreage estimate is usually a more accurate and precise estimate than that of the wetland delineator alone. Usually, the surveyed estimate is smaller in acreage than the relatively crude estimate of the delineator. A more accurate and precise boundary is a better product for the landowner and the USACE, and oftentimes the acreage estimate is smaller. This may be a vital consideration when the initial estimate of acreage exceeds 1 acre or some other critical quantity used in the permitting process.

Obtaining a survey is usually a simple matter because civil engineers and surveyors concerned with the design of the site plan are in possession of requisite knowledge and equipment to complete the mapping task (Figure 5.12). The services of surveyors can be made available to permanently record the wetland boundary delineated by the expert. The boundary or topographic mapping data used in the engineering site plan can be employed as a base map, and the wetland boundary placed as an overlay on the map.

The location of wetlands must be "flagged" in the field for the survey crew by personnel familiar with the extent of jurisdictional wetlands on site. It is desirable to have the wetland expert delineate the boundary and flag it and to have surveying personnel measure the area and map the boundary on the available site maps. It is also desirable to have the wetland expert accompany the surveyors to the site and familiarize them with the wetland boundaries and method of flagging. Upon completion of the survey, a wetland map and an estimate of wetland acreage can be made by surveyors and supplied to the delineator for use in the wetland report.

To provide an optimal survey of wetland boundaries for the purposes of permitting and to provide suitable detail for maps at scales of 1 in. = 50 ft to 1 in. = 200 ft, it is necessary to be respective of scale considerations in flagging and surveying the boundary. To obtain a good product, it is important for the wetland field personnel to overly "flag" the boundaries. It is also important to select a straight "line of sight" for distances of 5 to 10 ft or more.

An optimal amount would be 50 or more feet, though the variability in wetland shape may not allow this. The distance of 20 to 50 ft on the ground equates to a distance of 0.04 ft or 0.5 in. on a map of scale 1 in. = 100 ft. Hence, it serves a practical purpose to flag a boundary for the survey that includes relatively long, straight distances. It is also important that the surveyed and flagged boundary remain a true representation of the boundary of jurisdictional wetlands.

Figure 5.12 A survey stake in the foreground marks the end of a 100-ft increment of the survey baseline. Other markers are spaced 100 ft apart off into the distance. Survey markers and surveyed wetland boundaries can be of great assistance in wetland identification and permitting.

In the field, the wetland expert or other knowledgeable person can use a hand compass to help "line up" the "flags" and create straight boundary lines. The wetland personnel may also wish to pace these distances and record their angles using a compass. These data can be used to make a map for a wetlands report or to provide a preliminary map and assist surveyors in completing a more detailed survey. The existence of high frequency "flagging" and a general map will greatly facilitate the work of the surveyors and ensure an accurate and precise boundary.

It may also be desirable for the wetland expert to accompany the surveyors, in order to assure the quality and to resolve any boundary ambiguities on the spot. In the case of a lawsuit, this supervision of the survey would be vital.

This approach assures a high quality mapping product and an accurate and precise estimate of wetland area. This is a particularly important detail, as a well defined and mapped boundary is generally a correct estimate and produces the best result for the USACE. Such a map of large wetland areas is also greatly appreciated by government regulators.

6

Discussion

A variety of groups are interested in wetlands. These groups tend to focus on different elements of the issue. Some are concerned with hunting wildlife in wetlands, some with watching wildlife and plants, and others with conserving and preserving wildlife and wetlands. Among these groups there are both consumptive users and nonconsumptive users of wetland resources. Each group and its focus deserve attention. Each group has a valid interest in the use of the resource. Many of the groups have preserved much of the habitat we now have. We enjoy their foresight and should honor their contributions.

Naturally, this mix of interests and uses of wetland resources results in competition and controversy. In this sea of competitive interests and variety of opinions, the work of identification and delineation of wetlands must be conducted. The delineator must function as per instructions and guidance from the scientific and engineering community and use the regulatory guidance provided by USACE and any other agencies involved. The delineator must function in an objective fashion and produce a delineation and report that meets the needs of government entities who administer wetland regulations and permitting activities. It is also important to best represent the needs and rights of the landowner or client in developing the report and presenting results for inspection.

In this atmosphere of competitive interests and controversy, one must separate the fact from fiction and render a scientifically defensible, jurisdictional boundary. Therein lies many a challenge.

Wetland identification and delineation is a very difficult process. The process involves knowledge of science and natural history and requires knowledge of wetland and upland plants, soils, and hydrology (Figure 6.1). Delineation is necessarily a hard task for individuals, due to the variety of knowledge required to perform a complete analysis and report.

Knowledge of the various elements that make up a wetland and that are the basis of wetland delineation is difficult to master by one individual. This is probably due to the multiple disciplinary nature of characteristics that make up wetlands and the variety of functions wetlands perform. Often this problem is addressed when delineation efforts are conducted by a team of individuals. Each member is an expert in one or more of the wetland criterion. The team is often composed of a person knowledgeable in soils, another person knowledgeable in plants, and perhaps a third person in charge and knowledgeable in all three criterion of wetland conditions.

Hopefully, the material presented here will provide helpful ideas and techniques for delineation, as well as insight into the process. These techniques, combined with individual or team skills and experience, will assist in the development of familiarity with the delineation process. Through learning, training, and practice, the delineator can render the best identification of wetlands. A detailed, high quality wetland report will serve an important need for information and can potentially assist a variety of groups interested in wetlands.

THE PERMITTING PROCESS

Once the wetland delineation is made and its conclusions are known, it is possible to evaluate the need for a permit. Currently, whenever jurisdictional wetlands are present on a property, and the intent of the owner is to fill some of those wetlands, it is desirable to obtain a wetland delineation and report and then to determine whether a permit is necessary.

In general terms, the current situation necessitates applying for a permit when jurisdictional wetlands are to be filled. This is the case for wetlands that constitute the waters of the U.S.

Figure 6.1 The juxtaposition of the three wetland criterion is pictured for Colorado River wetlands north of Topock, Arizona.

A small fill is allowed under Nationwide Permit 26. The conditions allow up to 1 acre of fill, if the area is above the headwaters of the water drainage system. These areas are known as isolated wetlands. This also assumes that the fill does not violate other environmental conditions, and the State has approved the Nationwide Permit. Above the headwaters can be defined as being above the portion of the drainage that supports runoff exceeding 5 ft^3/sec. There may also be other definitions.

If a wetland is found in tidewater areas, if it contains unusual plant or animal species, or if it is below the headwaters of the watershed, the above condition changes. Then a permit is required for a fill or discharge of less than 1 acre.

Once the need for a permit is determined, the USACE can be contacted. This contact can be best made through the USACE District that has jurisdiction for the area under study. The best contact is the Regulatory Functions Branch of the District. District addresses are listed in Appendix A.

The initiation of discussion with the Regulatory Functions Branch will require information. This entails conversations or correspondence related to the presence of wetlands on the property and the intentions of the landowners concerning any change in land cover that departs from the current, wetland condition. Hence, a necessary part of the permit process is a wetland delineation and report, so as to be able to discuss the wetland conditions from a position of knowledge.

As part of this permit process, the USACE will probably visit the property. This can be accomplished in company with the landowner and/or the author of the wetland delineation report and perhaps an attorney for the landowner. The USACE will examine the condition of the property, look for wetland indicators, and provide information related to permitting.

If the wetland report is available, they will examine the adequacy and quality of the report and delineation. If the report is available, they can either agree with the wetland delineation and report or will suggest revisions that are deemed necessary.

After field visits, a revised delineation report and permit application can be submitted for review. If the revised report largely represents the consensus of the USACE and the landowner, as presented by the landowner or a designated agent, a Nationwide Permit may be issued for fills of 1 acre or less of isolated wetlands.

The simplest solution to the permitting matter is to seek consensus with the USACE concerning the presence and location of jurisdictional wetlands. There are few, simple remedies for a disagreement between the landowner and the USACE, though they are discussed in Want (1992) and other sources. To save time and achieve a good resolution, it is best to iterate to a solution agreeable to the landowner and to the USACE.

Often times, the findings of the USACE and the delineator are more than 1 acre of jurisdictional wetlands. A very good solution to an impasse on location and quantity of wetlands is to avoid disturbing the wetland resource. This is accomplished by redesign of the site plan to avoid or reduce fill of jurisdictional wetlands. This allows the landowner to avoid the issue of mitigating filled wetlands by reducing or eliminating the fill of jurisdictional wetlands. The action also necessitates protection of the wetland resource from impacts related to adjacent developments.

Commonly, the optimal course of action is to build around a wetland. This necessitates development of an engineering plan after the location and size of the wetlands have been determined.

Again, there is a need for a wetland delineation and report to guide the planning of the site or management of the resource. This effort should be completed early in the process of property acquisition and development, and if possible, it should proceed the engineering design phase of the project.

The alternative to this practice is to delineate and then fill the existing wetland (Figure 6.2). The current federal requirement is then to mitigate the action by constructing new wetlands on a less "central" part of the site. The wetlands can only be legally removed after the permitting process is completed and the mitigation approach or plan has been accepted by the USACE and other interested agencies.

The practice of mitigation through constructed wetlands is both costly and controversial. Mitigation projects are accompanied by many requirements in topical areas such as wetland

Figure 6.2 The practice of filling wetlands was common years ago as pictured in 1985 near Long Point, Lake Erie, Ontario, Canada.

characteristics and functions, permitting activities, and legal issues. One requirement is that similar habitat in function and quality be built and maintained over a long period of time. Mitigation is a serious exercise in engineering and wetland management.

Mitigation also creates a long-term liability for the owner as they must plan, permit, build, and maintain the constructed wetland habitat. This maintenance requirement is a potentially open liability.

An initial "milestone" used to evaluate quality of mitigation effort is that the constructed wetland display a variety of wetland "functions" that were present in the original, mitigated wetlands. Moreover, the function should be maintained over a long period, say 5 years or more. The implication is that it should function a long time, as the habitat it replaced was the result of many years of certain wetland conditions. This is another real liability to the landowner, and the cost implications are part of the problem of mitigating a filled wetland area. Hence, many constructed or mitigation wetlands are given to a public agency or other group for maintenance purposes.

There is a great controversy concerning whether mitigation wetlands have succeeded in duplicating or approximating the original conditions of the mitigated wetlands. This is a continuing source of discussion, and the entire process is subject to future change. This change will probably be due to the controversy and due to increased research yielding detailed findings as to success based on functional characteristics and duration of function of constructed wetlands.

Mitigation is sometimes necessary when the presence of the wetland will hinder use of the property. The issue of mitigation of wetlands is a difficult and complex one. The scope is beyond this effort. Interested parties can refer to other books on the subject for guidance (Hammer, 1989; Kusler and Kentula, 1989; Mitsch and Jorgensen, 1989; Salvensen, 1990; Hammer, 1991; Marble, 1992).

Clearly, the effort that can be involved in mitigation is great, and the potential costs can be high. It is more expedient and simpler to avoid the jurisdictional wetlands and build adjacent to them or at some other location. This will preserve the re-source and save time and money in the long term.

Either the practice of avoiding wetlands or filling wetlands and/or mitigating the damage requires a wetland assessment to be conducted prior to development (Figure 6.3). It is an important first step and part of the work to be completed in any site planning or development effort. Much as a developer, planner, or engineer would evaluate a site for buried gasoline tanks, hazardous wastes, flood plain areas, land ownership and titles, and other conditions that could interfere with development, each site must be evaluated for wetland resources.

ENVIRONMENTAL ATTORNEYS

Many questions related to wetlands are mostly legal questions. The people involved in delineation understand laws and regulations from a lay viewpoint and are able to comply with the applicable legal requirements. As is the case in many legal matters related to wetlands, it is often necessary to seek the assistance of a specialist or environmental attorney.

A Section 404 wetland permit evaluation by the USACE is designed to be done by the landowner, a designated engineer, or a wetland expert working for the landowner. However, the complexities of the law, government administration, science, and engineering are such that it is very wise to involve attorneys familiar with wetland issues.

From the landowners perspective, the potential costs associated with a work stoppage via cease and desist order argue for addressing wetlands in a direct, highly organized fashion. Errors can be costly, and all parties involved wish to avoid errors. Many of the errors can be related to legal questions. Environmental attorneys can often solve problems associated with the USACE permitting and focus resources on answering the questions in a germane and precise fashion.

It is also good to note that a scientific and engineering-based delineation and wetland report is a necessary first step in wetland analysis, and an attorney will eventually require this information for guidance purposes and for forming replies to the USACE questions related to permits.

Figure 6.3 Wetlands serve an important purpose as the interphase between aquatic and terrestrial environments. Much effort has been devoted to avoiding wetland areas for home building activities. This is particularly true of coastal zone areas. Old homes are potentially subject to natural hazards such as these destroyed during a December 1985 storm off of Lake Erie. Pictured is Long Point, Ontario, Canada.

THE POTENTIAL FOR CHANGE

The future of wetland delineation is very uncertain. Change is perhaps the only certainty.

There are many unsolved questions with regard to wetland issues. This stems from the recent attention devoted to wetlands and the resulting controversy. Further problems have resulted from the general history of limited interest in and limited research funding for wetland-related topics.

Questions such as those of wetland definition or questions where there are no current consensus may be clouded by this controversy (Figure 6.4). These questions are legitimate, and many of the questions relate to topical areas that need further scientific and engineering evaluations (Figure 6.5).

A number of issues need to be addressed in a direct, research-oriented fashion. The foundations of wetland delineation are based on certain characteristics and generalities of wetland conditions. These assumptions are very good. However, there are many exceptions to the rules, due to the variability of local conditions. Hence, research is necessary to address the general conditions and place them in a more rigorous context.

The special exceptions or special cases must also be addressed to assure that the regulations are administered correctly and uniformly. It is also important that wetland resources of special or rare conditions be maintained. This is true regardless of size.

Wetland issues need to be addressed in a more formal sense. They also need to be addressed for a wide variety of wetland types and the climatic and ecological regions they are found.

The list of questions is long, and a few are mentioned here. What is a wetland (Figure 6.6)? This is still an issue and certainly can be subject to change. Various groups are working on modifications or outright changes.

What is the hydrology of a wetland? Does the wetland hydrology definition include lands that are flooded or inundated 7, 14, or 21 days during the growing season? Should we consider time limits for surface flooding and for subsurface soil saturation or a combination of both?

Figure 6.4 Is this a jurisdictional wetland? Tundra and taiga areas of Alaska are underlined by permafrost, which ponds water and contributed melt water to the thin soil horizon unfrozen during the summer season. These areas may demonstrate all three criterion, as do millions of acres of interior Alaska. This presents a problem. Pictured is alpine tundra of the headwaters of the major Kobuk River, in the Arrigetch Peaks of the Gates of the Arctic National Park, Brooks Range.

Figure 6.5   Is this a jurisdictional wetland? It displays all three criterion and hence is jurisdictional. It is also a temporary drainage ditch and was never meant to "become" a wetland.

Figure 6.6 Is this stream running through a plowed field a jurisdictional wetland? Some would say so, and others would disagree. Some of the wetland criterion are present, but some are absent due to farming activities conducted on this South Dakota field.

What is the importance of ephemeral or temporary wetlands? In arid or frigid areas, wetlands are temporary in their presence and may not demonstrate the characteristics of jurisdictional wetlands on given occasions such as the day of a wetland delineation or wetland inspection (Figure 6.7).

How can ephemeral wetlands be delineated? If the land does not normally demonstrate the wetland criteria, how can we determine the presence and extent of wetland conditions?

What is a farmed wetland? If an area demonstrates most or all of the jurisdictional criterion and is covered with a crop, is it a jurisdictional wetland? Currently, it is a jurisdictional wetland, but will this change?

How can wetland losses be mitigated? How can we specify the desired wetland functions a constructed wetland should exhibit? How can we measure these functions of the original wetland before filling? How can we measure these functions over the many years that constructed wetlands will duplicate the functions of original wetlands?

Amidst this backdrop of uncertainty, it is incumbent upon the delineator to remain knowledgeable on current regulations. This includes regulatory guidance information related to permitting provided by the USACE. The delineator may or may not be involved in wetland mitigation, but should also be aware of the general characteristics and requirements of such an effort.

The need to remain knowledgeable can be met through a number of approaches to learning. These include obtaining copies of recent communications found in the Federal Register, maintaining one's address on lists for regulatory guidance newsletters issued by Public Affairs Offices or Regulatory Functions Branches of the USACE Districts, reading national and local newspapers, maintaining contact with environmental attorneys, and consulting professional journals. A summary and bibliography of Regulatory Guidance Letters can be found in the January 22, 1991 Federal Register.

Good sources can include materials published by the Society of Wetland Scientists (SWS, Post Office Box 296, Wilmington, NC 28402), the National Association of Wetland Managers (NAWM, Post Office Box 2463, Berne, NY 12023-9746), law journals, and/or newsletters such as the National Wetlands

Figure 6.7 This river flows on a temporary basis, based on storm hydrology. Though temporary, it is a vital resource as a riparian rangeland. Is it a jurisdictional wetland? Probably, it is a wetland in function during wet times of the year, but not at other times.

Newsletter (Environmental Law Institute, 1616 P Street NW, Washington, DC, 20036).

There are some fine books on a variety of issues related to wetlands. Depending on one's interest, several books or publications should be consulted (Martin et al., 1955; Shaw and Fredine, 1956; Ponnaperuma, 1972; Hutchinson, 1975; Wetzel, 1975; USDA, 1975; Teskey and Hinckley, 1978; Jorgensen and Mitsch, 1983; Gillen, 1986; Mitsch and Gosselink, 1986; Goldfarb, 1988; Mitsch et al., 1988; Niering, 1988; USDI, 1989; Foth, 1990; Gosselink et al., 1990; Wetzel and Likens, 1990; Want, 1992; USACE, 1992a, USACE, 1992b, to name a few).

A number of additional sources are presented in the Bibliography and in Appendix A on sources of data.

CONCLUSIONS

This book presents the background and methods necessary to complete an identification and delineation of jurisdictional wetlands. It provides insight on wetlands, their jurisdictional definition, and methods to better define their presence and incorporate this information in a wetland report or site plan. The current federal delineation regulations or manual (e.g., USACE, 1987), the USDA List of Hydric Soils (USDA, 1991), the federal list of wetland plants (Reed, 1988), as well as this book should be of value to the reader in completing an effort. The combination of these sources may also help the reader supervise a contractor to complete such a report or help the reader interpret a wetland delineation report.

Bibliography and Other Wetland Literature

BIBLIOGRAPHY OF GENERAL WETLAND AND WETLAND RELATED LITERATURE

Anderson, J., E. Hardy, J. Roach, and R. Witmer, 1976. A land use classification system for use with Remote-sensor data. U.S. Department of Interior, U.S. Geological Survey Professional Paper 964, Washington, D.C., 28 pp.

Cowardin, L., V. Carter, F. Golet, and E. LaRoe, 1979. Classification of wetlands and deepwater habitats of the United States. U.S. Department of Interior, U.S. Fish and Wildlife Service, Rep. No. FWS/OBS-79/31, Washington, D.C., 103 pp.

Federal Interagency Committee for Wetlands Delineation, 1989. Federal Manual for Identifying and Delineating Jurisdictional Wetlands. January 10, U.S. Government Printing Office, Washington, D.C., 76 pp.

Foth, J., 1990. *Fundamentals of Soil Science.* John Wiley, New York.

Gillen, L., 1986. *Photographs and Maps Go to Court.* American Society for Photogrammetry and Remote Sensing, Bethesda, MD, 88 pp.

Goldfarb, W., 1988. *Water Law.* Lewis Publishers, Chelsea, MI, 290 pp.

Good, R., D. Whigham, and R. Simpson, 1978. *Freshwater Wetlands: Ecological Processes and Management Potential.* Academic Press, New York, 380 pp.

Gosselink, J., L. Lee, and T. Muir, 1990. *Ecological Processes and Cumulative Impacts.* Lewis Publishers, Chelsea, MI, 708 pp.

Hammer, D., 1989. *Constructed Wetlands for Wastewater Treatment: Municipal, Industrial and Agricultural.* Lewis Publishers, Chelsea, MI, 800 pp.

Hammer, D., 1992. *Creating Freshwater Wetlands.* Lewis Publishers, Chelsea, MI, 298 pp.

Hutchinson, G., 1975. *A Treatise on Limnology, Vol. III, Limnological Botany.* John Wiley, New York, 660 pp.

Johnson, M. and W. Goran, 1987. Sources of digital spatial data for geographic Information Systems. U.S. Army Corps of Engineers, Construction Engineering Research Laboratory, Tech. Rep. No. N-88-01, Champaign, IL, 33 pp.

Jorgensen, S. and W. Mitsch, 1983. *Application of Ecological Modeling in Environmental Management.* Elsevier, New York.

Kusler, J. and M. Kentula, 1990. *Wetland Creation and Restoration.* Island Press, Covelo, CA, 596 pp.

Lee, F., E. Bentley, and R. Amundson, 1975. Effects of marshes on water quality. In *Coupling of Land and Water Systems.* A. Hassler, Ed. Springer-Verlag, New York, 309 pp.

Lee, K., 1991. Wetland detection methods investigation. U.S. Environmental Protection Agency, Environmental Monitoring Systems Laboratory, Rep. No. 600/4-91/014, Las Vegas, NV, 73 pp.

Lyon, J., 1978. An analysis of vegetation communities in the Lower Columbia River Basin. Proceedings of the Pecora Symposium on Applications of Remote Sensing to Wildlife Management, Sioux Falls, SD, pp. 321–327.

Lyon, J., 1979. Remote sensing of coastal wetlands and habitat quality of the St. Clair Flats, Michigan. Proceedings of the 13th International Symposium on Remote Sensing of Environment, Ann Arbor, MI, pp. 1117–1129.

Lyon, J., 1980. Data sources for analyses of Great Lakes wetlands. Proceedings of the Annual Meeting of the American Society for Photogrammetry, St. Louis, MO, pp. 512–525.

Lyon, J., 1981. The Influence of Lake Michigan Water Levels on Wetland Soils and Distribution of Plants in the Straits of Mackinac, Michigan. Doctoral Dissertation, University of Michigan, Ann Arbor, 155 pp.

Lyon, J., 1983. Landsat-derived land cover classification for locating potential Kestrel nesting habitat. *Photogramm. Eng. Remote Sensing* 49:245–250.

Lyon, J., 1987. Maps, aerial photographs and remote sensor data for practical evaluations of hazardous waste sites. *Photogramm. Eng. Remote Sensing* 53:515–519.

Lyon, J., 1989. Remote sensing of suspended sediments and wetlands in Western Lake Erie, in Lake Erie Estuarine Systems: Issues, Resources, Status, and Management. Kreiger, K., Ed. NOAA Estuary-of-the Month Seminar Series, Publ. No. 14, U.S. Department of Commerce, Washington, D.C., pp. 124–147.

Lyon, J. and R. Drobney, 1984. Lake level effects as measured from aerial photos. *J. Surv. Eng.* 110:103–111.

Lyon, J., K. Bedford, J. Chien-Ching, D. Lee, and D. Mark, 1988. Suspended sediment concentrations as measured from multidate Landsat and AVHRR data. *Remote Sensing Environ.* 25:107–115.

Lyon, J., R. Drobney, and C. Olson, 1986a. Effects of Lake Michigan water levels on wetland soil chemistry and distribution of plants in the Straits of Mackinac. *J. Great Lakes Res.* 12:175–183.

Lyon, J. and R. Greene, 1992. Lake Erie water level effects on wetlands as measured from aerial photographs. *Photogramm. Eng. Remote Sensing,* 58:1355–1360.

Lyon, J., J. Heinen, R. Mead, and N. Roller, 1987. Spatial data for modeling wildlife habitat. *ASCE J. Surv. Eng.* 113:88–100.

Lyon, J., R. Lunetta, and D. Williams, 1992. Airborne multispectral scanner data for evaluation of bottom types and water depths of the St. Marys River, Michigan. *Photogramm. Eng.Remote Sensing* 58:951–956.

Lyon, J., J. McCarthy, and J. Heinen, 1986b. Video digitization of aerial photographs for measurement of wind erosion damage on converted rangeland. *Photogramm. Eng. Remote Sensing* 52:373–377.

Lyon, J. and C. Olson, 1983. *Inventory of Coastal Wetlands.* Michigan Sea Grant Program Publ., University of Michigan, Ann Arbor, 35 pp.

Marble, A., 1992. *A Guide to Wetland Functional Design.* Lewis Publishers, Chelsea, MI, 222 pp.

Martin, A, H. Zim, and A. Nelson, 1955. *American Wildlife and Plants*. Dover Publications, New York, 500 pp.

Mitsch, W. and J. Gosselink, 1986. *Wetlands*. Van Nostrand Reinhold, New York, 539 pp.

Mitsch, W. and S. Jorgensen, 1989. *Ecological Engineering, and Introduction to Ecotechnology*. John Wiley, New York, 472 pp.

Mitsch, W., M. Straskraba, and S. Jorgensen, 1988. *Wetland Modelling*. Elsevier, New York.

Mortimer, C., 1941. The exchange of dissolved substances between mud and water in lakes. *J. Ecol.* 29:280–329.

Mortimer, C., 1971. Chemical exchanges between sediments and water in the Great Lakes — speculation on probable regulatory mechanisms. *Limnol. Oceanogr.* 16:387–404.

Mudroch, A. and S. MacKnight, 1991. *CRC Handbook of Techniques for Aquatic Sediments Sampling*. CRC Press, Boca Raton, FL, 208 pp.

Mueller-Dombois, D. and H. Ellenburg, 1974. *Aims and Methods of Vegetation Ecology*. John Wiley, New York, 547 pp.

Munsell Color, 1990. *Munsell Soil Color Charts*. Kollmorgen Instruments Corporation, Baltimore, MD, 18 pp.

Napier, T., 1990. *Implementing the Conservation Title of the Food Security Act of 1985*. Soil and Water Conservation Society, Ankeny, IA, 363 pp.

Niering, W., 1988. Wetlands. *Audubon Society Nature Guide*. A. Knopf, New York, 638 pp.

Ormsby, J. and R. Lunetta, 1987. Whitetail deer food availability maps from thematic mapper data. *Photogramm. Eng. Remote Sensing* 53:1585–1589.

Ponnamperuma, F., 1972. The chemistry of submerged soils. *Adv. Agron.* 24:29–96.

Reed, P., 1988. National list of plant species that occur in wetlands: national summary. U.S. Department of Interior, U.S. Fish and Wildlife Service, Biological Rep. 88 (24), 244 pp.

Roller, N., 1977. *Remote Sensing of Wetlands*. Environmental Research Institute of Michigan, Ann Arbor, MI.

Salvesen, D., 1990. *Wetlands: Mitigating and Regulating Development Impacts*. Urban Land Institute, Washington, D.C.

Sculthorpe, C., 1967. *The Biology of Aquatic Vascular Plants*. St. Martin Press, New York, 609 pp.

Shaw, S. and C. Fredine, 1956. Wetlands of the United States. U.S. Fish and Wildlife Service, Circular 39, 67 pp.

Teskey, R. and T. Hinckley, 1978. Impact of water level changes on woody riparian and wetland communities. U.S. Fish and Wildlife Service, Rep. No. FWS/OBS-78/89, Volumes I through VI, Washington, D.C.

Want, W., 1992. *Law of Wetlands Regulation.* Clark, Boardman and Callaghan, Deerfield, IL.

Wetzel, R., 1975. *Limnology.* Saunders, Philadelphia, 743 pp.

Wetzel, R. and G. Likens, 1990. *Limnological Analyses.* Springer-Verlag, New York, 368 pp.

Williams, D. and J. Lyon, 1991. Use of a Geographical Information System data base to measure and evaluate wetland changes in the St. Marys River, Michigan. *Hydrobiologia* 219:83-95.

U.S. Army Corps of Engineers, 1987. Corps of Engineers Wetlands Delineation Manual. *Tech. Rep.* Y-87-1, Department of the Army, Washington, D.C.

U.S. Army Corps of Engineers, 1992a. Regulatory permit program. Great Lakes Levels, update letter no. 79, North Central Division, Chicago, IL, February 3, 4 pp.

U.S. Army Corps of Engineers, 1992b. *Photogrammetric Mapping.* Engineering Manual, Washington, D.C.

U.S. Department of Agriculture, 1962. *Soil Survey Manual.* Soil Conservation Service, Washington, D.C., 502 pp.

U.S. Department of Agriculture, 1975. *Soil Taxonomy, A Basic System of Soil Classification for Making and Interpreting Soil Surveys.* U.S. Soil Conservation Service, Washington, D.C., 754 pp.

U.S. Department of Agriculture, 1988. *National Food Security Act Manual.* U.S. Soil Conservation Service, Washington, D.C.

U.S. Department of Agriculture, 1991. Hydric soils of the United States. Misc. Publ. 1491, Soil Conservation Service, Washington, D.C., 502 pp.

U.S. Department of Interior, 1989. *Flood Hydrology Manual.* Bureau of Reclamation, Washington, D.C., 243 pp.

U.S. Environmental Protection Agency, 1991. Federal manual for identifying and delineating jurisdictional wetlands. *Fed. Reg.*, August 14.

BIBLIOGRAPHY OF PLANT RELATED LITERATURE

The following references have been found to be useful for identification of upland and wetland plants. They range in content from books covering the regional types and distribution of plants to books that address national or North American distributions. They also range in the level of detail they supply in describing the plants. All of these books will be potentially helpful in delineation when used along with local books on plants.

Agricultural Research Service, 1971. *Common Weeds of the United States*. U.S. Department of Agriculture, Washington, D.C., and Dover Publ., New York, 463 pp.

Billington, C., 1968. *Shrubs of Michigan*. Cranbrook Institute for Science, Bloomfield Hill, MI, 339 pp.

Braun, E. L., 1964. *Deciduous Forests of Eastern North America*. Hafner, New York, 596 pp.

Braun, E. L., 1989. *The Woody Plants of Ohio*. Ohio State University Press, Columbus, 362 pp.

Brown, L., 1979. Grasses, an Identification Guide. Houghton Mifflin, Boston, 240 pp.

Britton, N. and A. Brown, 1970. *An Illustrated Flora of the Northern United States and Canada*. Volumes I, II, and III, Dover Publ., New York, 2052 pp.

Cobb, B., 1963. *Ferns*. Houghton Mifflin, Boston, 281 pp.

Courtenay, B. and J. Zimmerman, 1972. *Wildflowers and Weeds*. Van Nostrand Reinhold Co., New York, 144 pp.

Dana, W., 1963. *How to Know the Wild Flowers*. Dover Publ., New York, 418 pp.

Eleuterius, L., 1990. Tidal Marsh Plants. Pelican Publ., Gretna, LA, 168 pp.

Fassett, N., 1957. *A Manual of Aquatic Plants*. University of Wisconsin Press, Madison, 339 pp.

Fassett, N., 1978. *Spring Flora of Wisconsin*. University of Wisconsin Press, Madison, 413 pp.

Fisher, T., 1988. *The Dicotyledoneae of Ohio, Part III, Asteraceae*. Ohio State University Press, Columbus, 280 pp.

Forey, P., 1990. *Wildflowers.* Gallery Books, New York, 239 pp.

Harlow, W., 1946. *Fruit Key and Twig Key.* Dover Publ., New York, 126 pp.

Harlow, W., 1957. *Trees of the Eastern and Central United States and Canada.* Dover Publ., New York, 288 pp.

Haslam, S., 1978. River Plants. Cambridge University Press, Cambridge, UK, 396 pp.

Hitchcock, A., 1971. *Manual of the Grasses of the United States, Volumes I and II.* Dover Publ., New York, 1617 pp.

Hotchkiss, N., 1972. *Common Marsh, Underwater and Floating-leaved Plants of the United States and Canada.* Dover Publ., New York, 124 pp.

Knobel, E., 1977. *Field Guide to the Grasses, Sedges and Rushes of the United States.* Dover Publ., New York, 83 pp.

Knobel, E., 1988. *Identify Trees and Shrubs by their Leaves, a Guide to Trees and Shrubs Native to the Northeast.* Dover Publ., New York, 47 pp.

Little, E., 1979. *Forest Trees of the United States and Canada, and How to Identify Them.* Dover Publ., New York, 70 pp.

Little, E., 1980. *The Audubon Society Field Guide to North American Trees, Eastern Region.* A. Knopf, New York, 714 pp.

Martin, E., 1984. *A Beginner's Guide to Wildflowers of the C and O Towpath.* Smithsonian Institution, Washington, D.C., 72 pp.

Miller, H. and S. Lamb, 1985. *Oaks of North America.* Naturegraph Publ., Happy Camp, CA, 327 pp.

Mohlenbrock, R., 1987. *Wildflowers, a Quick Identification Guide to the Wildflowers of North America.* Macmillan, New York, 203 pp.

Mohlenbrock, R. and J. Thieret, 1987. *Trees, a Quick Reference Guide to Trees of North America.* Macmillan, New York, 155 pp.

Muensoher, W., 1980. Weeds. Second Edition. Cornell University Press, Cornell, NY, 586 pp.

Niering, W. and N. Olmstead, 1979. *The Audubon Society Field Guide to North American Wildflowers, Eastern Region.* A. Knopf, New York , 886 pp.

Peterson, R. and M. McKenny, 1968. *A Field Guide to Wildflowers of Northeastern and North-central North America.* Houghton Mifflin, Boston, 420 pp.

Petrides, G., 1988. *A Field Guide to Eastern Trees, Eastern U.S. and North America*. Houghton Mifflin, Boston, 272 pp.

Preston, R., 1989. *North American Trees*. Iowa State University Press, Ames, 407 pp.

Sargent, C., 1965. *Manual of the Trees of North America*, Volumes I and II. Dover Publ., New York.

Smith, H., 1966. *Michigan Wildflowers*. Cranbrook Institute for Science, Bloomfield Hills, MI, 468 pp.

Spencer, E., 1957. *All About Weeds*. Dover Publ., New York, 333 pp.

Symonds, G., 1958. *The Tree Identification Book*. W. Morrow, New York, 272 pp.

Symonds, G., 1973. *The Shrub Identification Book*. W. Morrow, New York, 379 pp.

Trelease, W., 1967. *Winter Botany, an Identification Guide to Native Trees and Shrubs*. Dover Publ., New York, 83 pp.

U.S. Army Corps of Engineers, 1977. Wetland Plants of the Eastern United States. North Atlantic Division, NADP 200-1-1, New York, 45 pp.

U.S. Army Corps of Engineers, 1979. A Supplement to Wetland Plants of the Eastern United States. North Atlantic Division, NADP 200-1-1 Supplement 1, New York, 100 pp.

U.S. Department of Agriculture, 1971. Common Weed of the United States. Dover Publ., New York, 463 pp.

U.S. Department of Agriculture, 1973. *Silvicultural Systems for the Major Forest Types of the United States*. U.S. Forest Service, Agriculture Handbook No. 445, Washington, D.C., 114 pp.

Voss, E., 1972. *Michigan Flora, Part I: Gymnosperms and Monocots*. Cranbrook Institute of Science, Bloomfield Hills, MI, 488 pp.

Appendix A

Sources of Data Including Aerial Photos, Maps, and Other Data Sources

SOURCES OF AERIAL PHOTOS

U.S. Geological Survey (USGS)

The EROS Data Center archives and produces copies of aerial photographs and other imagery acquired by Department of the Interior agencies. These acquisitions include USGS mapping photographs; products from high altitude aerial photography programs such as NAPP and NHAP; and products of different emulsion types such as color infrared, color, and black and white infrared; and aerial radar images. EDC also archives Landsat satellite data and NASA space and aerial images. A computerized data base of products can be accessed by latitude and longitude locators or by USGS quadrangle or other reference location. Computer printouts and microfiche of high altitude aerial photographs are available at no charge. Copies of products can be ordered from EDC.

For assistance, please contact: User Services, EROS Data Center, Sioux Falls, SD 57198, (605) 594-6151.

Agricultural Stabilization and Conservation Service (ASCS)

The ASCS holds photographs acquired by agencies of the Department of Agriculture. These groups include the U.S. Forest Service, U.S. Soil Conservation Service (SCS), the ASCS, and other agencies. Listings of holdings are supplied by computer printout and are generally available at no cost. One can access their holdings by latitude and longitude of the site or by county name.

For information, contact: Aerial Photography Field Office, ASCS-USDA, 2222 West 2300 South, POB 30010, Salt Lake City, UT 84125, (801) 975-3503.

National Archives and Record Service (NARS)

The NARS archives a variety of photo and map data and, in particular, aerial photographs flown by the government previous to 1942. Generally, one or two dates of aerial photography coverage are available for counties in the U.S. They publish a pamphlet that describes available coverage and provides ordering details.

For more information, please contact: National Archives and Records Service, Cartographic Branch, Washington, D.C., 20408.

National Ocean Survey (NOAA-NOS), Coastal Mapping Division

The Coastal Mapping Division acquires data over coastal and offshore areas in support of its mapping mandates.

For more information, contact: Coastal Mapping Division, National Oceanic and Atmospheric Administration (NOAA), Photographic Branch, N/CG 2314, Rockville, MD 20852, (301) 443-8601.

The efforts of the delineator involve federal lands themselves or land in-holdings found within federal management and ownership boundaries. In these cases, appropriate aerial photographs may be available locally or regionally from the following agencies.

US Forest Service (USFS)

The USFS contracts and acquires aerial photographic coverage over national forest lands and other areas related to their mandates. A variety of scales and film types have been employed. Most of these photographs are available through the USDA-ASCS Aerial Photography Field Office in Salt Lake City.

The USFS has nine regions with regional headquarters where regional aerial photographic coverage can often be viewed and ordered. Consult the government section of your telephone book for details.

Inquiries may also be referred to: Division of Engineering, U.S. Forest Service, Washington, D.C., 20250.

U.S. Bureau of Land Management (BLM)

The BLM has obtained coverage of the lands they manage. Local coverage is available through the USDA-ASCS Aerial Photography Field Office in Salt Lake City.

Inquiries may also be made to: U.S. BLM, Office of Public Affairs, Washington, D.C., 20240.

U.S. National Park Service (NPS)

The NPS contracts for aerial photography over and adjacent to U.S. national park lands and other areas such as national monuments and national recreational areas. Their holdings are available through the EROS Data Center in Sioux Falls.

Additional information may be available from: NPS, Office of Public Inquiries, Room 1013, Washington, D.C., 20240 or NPS, Denver Service Center, 655 Parfet Street, POB 25287, Denver, CO, 80225.

MAPS

Small scale topographic and other maps are made available by the U.S. Geological Survey. These maps may be purchased locally from vendors such as map stores, climbing and outdoor shops, and hunting and fishing stores. Naturally, these maps can be obtained from the USGS. One should be aware of the time frame necessary for delivery of products. It may be faster to obtain maps locally or to copy maps available in an archive such as a library or state natural resource agency.

Map orders may be placed through: USGS Map Sales, Box 25286, Denver, CO 80225.

The public can purchase maps on a "walk up" basis at USGS Earth Science Information Centers (ESIC). The USGS operates the larger ESIC, and they include the following:

Anchorage-ESIC
4230 University Drive, Room 101
Anchorage, AK 99508-4664

Anchorage-ESIC
Room G-84
605 West 4th Avenue
Anchorage, AK 99501

Denver-ESIC
169 Federal Building
1961 Stout Street
Denver, CO 80294

Lakewood-ESIC
Box 25046, Federal Center, MS 504
Building 25, Room 1813
Denver, CO 80225-0046

Menlo Park-ESIC
Building 3, MS 532, Room 3128
345 Middlefield Road
Menlo Park, CA 94025

Reston-ESIC
USGS
507 National Center
Reston, VA 22092

Rolla-ESIC
1400 Independence Road
Rolla, MO 65401

Salt Lake City-ESIC
8105 Federal Building
125 South State Street
Salt Lake City, UT 84138

San Francisco-ESIC
504 Custom House
555 Battery Street
San Francisco, CA 94111

Sioux Falls-ESIC
EROS Data Center
Sioux Falls, SD 57198

Spokane-ESIC
678 U.S. Courthouse
West 920 Riverside Avenue
Spokane, WA 99201

Stennis Space Center-ESIC
Building 3101
Stennis Space Center, MS 39529

Washington, D.C.-ESIC
U.S. Department of Interior
1849 C Street, NW, Room 2650
Washington, D.C. 20240

In all the states, there exists a local ESIC contact. These units are based at state natural resource agencies, state universities, and other agencies that have an interest in helping the public obtain maps and aerial photographs. Commonly, the local ESIC will maintain aerial photographs of the state, and they have access to USGS data bases of existing aerial photos. Hence, their assistance can be very valuable in identifying local aerial photo coverages.

The local ESIC offices often are very knowledgeable concerning aerial photographs held by state agencies and private companies. It may not be possible to access these local sources of photos by interaction with a regional or national data base.

The U.S. Geological Survey also vends a variety of digital cartographic and geographic data. These computer-compatible files allow the user to access data digitally and conduct computer processing and graphical display exercises. The most interesting data are the Digital Elevation Map (DEM) products, which are files of point elevations, and the Digital Line Graph (DLG) products, which display cultural or planimetric details such as roads and resource data such as drainage systems. Should the application require these sorts of information, DEM and DLG data sets can be of great assistance in regional analyses of wetlands (Lyon et al., 1987). You may wish to identify other sources of digital data. A good reference is that of Johnson and Goran (1987).

National Wetland Inventory Maps

National Wetland Inventory (NWI) Maps and information are available from: National Wetlands Inventory, 9720 Executive Center Drive, Suite 101, Monroe Building, St. Petersburg, FL 33702.

These wetland inventory maps are also available in digital form to facilitate computer processing and display. A very good use of such data would be in a Geographic Information System (GIS) data base. A sample of the utility of these data is provided by the 45-year study of wetlands along the St. Marys River area in Michigan (Williams and Lyon, 1992). These digital data are available from: National Wetlands Inventory Digital Cartographic and Geographic Data, U.S. Geological Survey, Earth Science Information Center (ESIC), 507 National Center, Reston, VA 22092.

Flood Plain Maps

Flood plain maps are available from the Federal Emergency Management Agency (FEMA) through its local, state, and regional offices. These offices can be located through the government section of the telephone directory. These maps and information can also be obtained from: FEMA, 500 C Street SW, Washington, D.C. 20472.

WEATHER INFORMATION

The National Weather Service of the Department of Commerce can be contacted for a variety of weather records. Products that record the weather conditions at a neighboring station are available on a daily basis or sometimes an hourly basis. Summary statistics by location, region, and state are also available. Many times these publications and general records are deposited in libraries and at universities, where they may be accessed quickly and at no charge.

For weather records, write to: National Weather Service, National Climatic Center, Federal Building, Ashville, NC 28801.

REGIONAL HYDROLOGICAL INFORMATION

The U.S. Geological Survey maintains offices in most states, and many of these offices are responsible for collecting hydrological information. The information may include water gaging station data for major and minor rivers and streams, and water quality sampling for selected locations.

The local USGS offices concerned with hydrology can be accessed through the government pages of the telephone directory. One can also contact the USGS directly at the larger centers listed under the map section and through the USGS national clearinghouses of information. One can also contact: Hydrologic Information Unit, U.S. Geological Survey, 419 National Center, Reston, VA 22092.

OTHER DATA SOURCES

The National Technical Information Service archives government reports and governmental contracted reports of science and engineering activities. It can be a great source of very detailed and technical information on a number of subjects.

Write to: National Technical Information Service, U.S. Department of Commerce, 5285 Port Royal Road, Springfield, VA 22161.

The U.S. Government Printing Office is a good source of current governmental publications. They operate a number of bookstores throughout the nation, and they are commonly found in federal buildings in larger cities. Locate the local bookstores through the government section of the telephone directory.

The Office may also be contacted at: Superintendent of Documents, U.S. Government Printing Office, North Capitol and H Streets, NW, Washington, D.C. 20402.

U.S. ARMY CORPS OF ENGINEERS COMMANDS

When corresponding with U.S. Army Corps of Engineers (USACE) Commands, it is best to locate the District level office that has regional jurisdiction in the area of interest. The District jurisdictions are often based on the boundaries of watersheds and may not be easy to ascertain. Use the local telephone directory and the government pages to locate the appropriate Command.

The field activities of the USACE are organized under Divisions, and the activities are conducted by the Districts. The initial point of contact for a wetland-related issue is the Regulatory Functions Branch of the District.

The USACE Districts and Divisions are listed here.

Alaska District
USACE
Bluff and Plum Street
Building 21-700, Room 200
Elmendorf AFB, AK 99506-0898

Albuquerque District
USACE
Box 1580
Albuquerque, NM 87103-1580

Baltimore District
USACE
Box 1715
Baltimore, MD 21203-1715

Buffalo District
USACE
1776 Niagara Street
Buffalo, NY 14207-3199

Chicago District
USACE
111 North Canal Street
Chicago, IL 60606-7206

Detroit District
USACE
Box 1027
Detroit, MI 48231-1027

Fort Worth District
USACE
819 Taylor Street
Box 173000
Fort Worth, TX 76102-0300

Honolulu District
USACE
Building 200
Fort Shafter, HI 96858-5440

Huntington District
USACE
502 Eighth Street
Huntington, WV 25701-2070

Jacksonville District
USACE
400 West Bay Street
Jacksonville, FL 32202-1191

Kansas City District
USACE
601 East 12th St.
Kansas City, MO 64106-2896

Little Rock District
USACE
Box 867
700 W. Capitol Avenue
Little Rock, AR 72203-0867

Los Angeles District
USACE
Box 2711
Los Angeles, CA 90053

Louisville District
USACE
600 Dr. Martin Luther King, Jr. Place
Louisville, KY 40202-2230

Memphis District
USACE
B-202, C. Davis Federal Building
Memphis, TN 38103-1894

Mobile District
USACE
109 St. Joseph Street
Mobile, AL 36602

New Orleans District
USACE
Box 60267
New Orleans, LA 70160

New York District
USACE
J. K. Javits Federal Building
New York, NY 10278-0090

Norfolk District
USACE
803 Front St.
Norfolk, VA 23510-1096

Omaha District
USACE
214 North 17th St.
Omaha, NE 68102-4978

Philadelphia District
USACE
100 Penn Square East
Philadelphia, PA 19107-3390

Pittsburgh District
USACE
1000 Liberty Ave.
Pittsburgh, PA 15222-4186

Portland District
USACE
Box 2946
Portland, OR 97208-2946

Rock Island District
USACE
Box 2004
Rock Island, IL 61204-2004

Sacramento District
USACE
1325 J St.
Sacramento, CA 95814-2922

San Francisco District
USACE
211 Main Street, Room 809
San Francisco, CA 94105-1905

Savannah District
USACE, Box 889
100 W. Oglethorpe
Savannah, GA 31401-0889

Seattle District
USACE
Box 3755
4735 East Marginal Way South
Seattle, WA 98124-2255

St. Louis District
USACE
1222 Spruce St.
St. Louis, MO 63103

St. Paul District
USACE
180 Kellogg Blvd. East, Room 1421
St. Paul, MN 55101-1479

Tulsa District
USACE
Box 61
Tulsa, OK 74121-0061

Vicksburg District
USACE
2101 N. Frontage Rd.
Vicksburg, MS 39180-5191

Walla Walla District
USACE
Bldg. 602, City-County Airport
Walla Walla, WA 99362-9265

Wilmington District
USACE
Box 1890
Wilmington, NC 28402-1890

Appendix B

Sample Intermediate Level Wetland Report and Data Records

The following is a sample wetland report from an intermediate level wetland delineation. The description and data record sheets follow the methods and recommendations made in the text of this book and in wetland manuals. The example is fictitious, and as such only a few intermediate level record sheets are included.

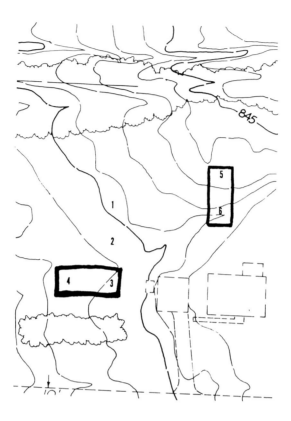

Large scale topographic map (1 in. to 100 ft) of the study site and sample locations.

The routine and intermediate level wetland evaluations used here have employed the techniques suggested in the government's, "Federal Manual for Identifying and Delineating Jurisdictional Wetlands," or referred to here as the Federal Interagency Wetland Manual of 1989. The conclusions derived here are consistent with and fully respective of jurisdictional wetland criteria recorded in the 1987 U.S. Army Corps of Engineers Wetlands Delineation Manual and the 1989 Federal Interagency Wetland Manual. Other sources were also evaluated to assist in this effort, including soil surveys, aerial photographs, plant identification books, and other wetland analysis references. These sources are listed in the Bibliography that follows.

Information from these sources and results from these analyses are described under the appropriate topic headings.

Soils

The Soil Survey for the county indicated that the site was composed of Bennington (Be) and Pewamo (Pm) soils. These Pewamo soils are on the national list of potential hydric soils. The Soil Survey map for the site is included here.

During the routine and intermediate level evaluations, I noted the soil types encountered in the field and their characteristics, and this information was utilized in my findings as per instructions in the Federal Interagency Wetland Manual (1989) and the U.S. Army Corps of Engineers Wetlands Delineation Manual (1987). The criteria set forth in the 1987 Manual was used in determining the presence/absence of hydric soils.

In particular, I evaluated the soils throughout the property during each field visit. This involved walking the entire site and examining the soil types, the drainage conditions, and indicators of hydric soils. This effort included probing of soils and observations of color, moisture, and indicators of hydric conditions.

Hydrology

I also noted the general hydrology of the site, and this information was utilized in my findings as per instructions in the Federal Interagency Wetland Manual (1989) and the U.S. Army Corps of Engineers Wetlands Delineation Manual (1987).

The site has no waterways such as a river, stream, or similar conveyance. The property is across the road and at considerably higher elevation than Crystal Creek. In general, it is flat with a cover of prior converted croplands and forest cover of 20 to 30 years in age formed on prior converted croplands. It appears that this site is above the headwaters of the watershed. This was judged by the absence of still or flowing water, the lack of drainage features on the USGS map, and the lack of adjacent stream or gully features.

The routine level evaluation of soils and hydric conditions identified six areas of potential wetland conditions. These areas demonstrated the presence of one or more indicators of wetland characteristics. These areas were later evaluated by intermediate level methods for soil characterization and by additional intermediate measures of hydrological and wetland plant characteristics. These intermediate sampling areas are noted on the enclosed map.

Plant Analyses

The site was visited during May 1992 to collect information on plants and other characteristics including soils and general hydrology of the area. This was done according to procedures described in the Federal Interagency Wetland Manual (1989) and the U.S. Army Corps of Engineers Wetlands Delineation Manual (1987).

The site was composed of wooded or former agricultural lands. These lands were examined on each field visit to observe their condition. Areas other than the potential wetland areas were not evaluated further because the routine evaluation demonstrated that: (1) the hydric soil indicators were absent, (2) the hydrological indicators were absent, and (3) wetland plants were infrequently encountered and certainly demonstrated no dominance on the site. Hence, these prior converted farmlands or forested areas exhibited no wetlands. In addition, evaluation of maps and aerial photos did not reveal potential wetland conditions at these locations.

The potential wetland areas were further evaluated using the intermediate level methods described in the Federal Interagency Wetland Manual (1989) and the U.S. Army Corps of Engineers Wetlands Delineation Manual (1987). The sampling locations are presented on the enclosed map. The sampling was limited to support of this reconnaissance effort.

Potential wetland locations were sampled and information for the worksheets was developed from: (1) field visits to identify and collect plants, and (2) from visual estimates of dominance and areal cover in the ground, shrub, and tree layer of vegetation. Results of these efforts are reported on a worksheet for each sampling location.

Plant areal estimates were based on the plants found within an approximate 10m (30 ft.) radius circle located at the sample grid point area. Estimates were made in increments as described in the Federal Interagency Wetland Manual (1989) and recorded in the field. All plant analyses were conducted when the plants were mature or in residue form. The results are indicative of the types of vegetation that the site will support. Plant species were keyed using a number of local and regional plant species identification books. The wetland determination calculations were then completed and recorded on the worksheets, and the results used to make the assessments.

The representative plants found on upland and wetland areas are described in detail on the intermediate level sampling sheets. In general, the adjacent wooded areas included plants such as Shagbark Hickory, Sugar Maple, and White Oak on upland sites. The adjacent wetter areas exhibited plants such as Pin Oak, Sugar Maple, and Red Elm. Prior converted farmland on adjacent upland locations demonstrated the presence of goldenrods, upland grasses, and other upland plants. Lowland areas of prior converted farmlands demonstrated the presence of cattail, rushes, bulrushes, and other wetland plants.

Wetland Mapping

Several areas of jurisdictional wetlands and potential jurisdictional wetlands were identified on the site. Locations of these areas are provided on the enclosed map, as is the general location of the study site. One area was found in the northeastern portion of the converted farmlands and was small at approximately 0.2 acres in size. A second area was found in the forested portion of the site on the southeastern boundary and was probably less than 0.3 acres in size. The total jurisdictional wetlands on the study site is potentially less than 0.5 acres.

DATA FORM

5/1/92

Field Investigator(s): _Dr. John G. Lyon_ Date: _____
Project/Site: _____ State: _OH_ County: _Franklin_
Applicant/Owner: _____ Plant Community #/Name: _____ [1]
Note: If a more detailed site description is necessary, use the back of data form or a field notebook.

- -

Do normal environmental conditions exist at the plant community?
Yes _X_ No _____ (If no, explain on back)
Has the vegetation, soils, and/or hydrology been significantly disturbed?
Yes _____ No _x_ (If yes, explain on back)

- -

VEGETATION

Dominant Plant Species	Indicator Status	Stratum	Dominant Plant Species	Indicator Status	Stratum
1. Daucus carota	UPL	Ground	11.		
2. Dipsacus fullonum	UPL	Ground	12.		
3. Solidago graminfolia	UPL	Ground	13.		
4. Rubus allegheniensis	FACU–	Ground	14.		
5. Juncus gerardi	FACW+	Ground	15.		
6.			16.		
7.			17.		
8.			18.		
9.			19.		
10.			20.		

Percent of dominant species that are OBL, FACW, and/or FAC _____20_____ %
Is the hydrophytic vegetation criterion met? Yes _____ No _X_
Rationale: _____

SOILS

Series/phase: _____Bennington_____ Subgroup:[2] _____
Is the soil on the hydric soils list? Yes _____ No _X_ Undetermined _____
Is the soil a Histosol? Yes _____ No _X_ Histic epipedon present? Yes _____ No _X_
Is the soil: Mottled? Yes _____ No _X_ Gleyed? Yes _____ No _X_
Matrix Color: _10YR 5/4, 10YR 5/4_ Mottle Colors: _____
Other hydric soil indicators: _____
Is the hydric soil criterion met? Yes _____ No _X_
Rationale: _____

HYDROLOGY

Is the ground surface inundated? Yes _____ No _X_ Surface water depth: _____
Is the soil saturated? Yes _____ No _X_
Depth to free-standing water in pit/soil probe hole: _____
List other field evidence of surface inundation or soil saturation.

Is the wetland hydrology criterion met? Yes _____ No _X_
Rationale: _____

JURISDICTIONAL DETERMINATION AND RATIONALE

Is the plant community a wetland? Yes _____ No _x_
Rationale for jurisdictional decision: _____

[1] This data form can be used for the Hydric Soil Assessment Procedure and the Plant Community Assessment Procedure.
[2] Classification according to "Soil Taxonomy."

DATA FORM

5/1/92

Field Investigator(s): __Dr. John G. Lyon__   Date: _____
Project/Site: _____ State: __OH__ County: __Franklin__
Applicant/Owner: _____ Plant Community #/Name: ___2___
Note: If a more detailed site description is necessary, use the back of data form or a field notebook.

- -

Do normal environmental conditions exist at the plant community?
Yes __X__ No _____ (If no, explain on back)
Has the vegetation, soils, and/or hydrology been significantly disturbed?
Yes _____ No __X__ (If yes, explain on back)

- -

VEGETATION

| | Indicator | | | | Indicator | |
Dominant Plant Species	Status	Stratum	Dominant Plant Species		Status	Stratum
1. Solidago graminfolia	UPL	Ground	11.			
2. Juncus gerardi	FACW+	Ground	12.			
3. Daucus carota	UPL	Ground	13.			
4. Aster pilosus	UPL	Ground	14.			
5.			15.			
6.			16.			
7.			17.			
8.			18.			
9.			19.			
10.			20.			

Percent of dominant species that are OBL, FACW, and/or FAC ____25____ %
Is the hydrophytic vegetation criterion met?   Yes _____ No _X_
Rationale: _____

SOILS

Series/phase: _____Bennington_____ Subgroup:[2] _____
Is the soil on the hydric soils list?   Yes _____ No _X_   Undetermined _____
Is the soil a Histosol? Yes _____ No _X_ Histic epipedon present? Yes _____ No _X_
Is the soil: Mottled? Yes _____ No _X_ Gleyed? Yes _____ No _X_
Matrix Color: __10YR 5/4, 10YR 5/4__ Mottle Colors: _____
Other hydric soil indicators: _____
Is the hydric soil criterion met?   Yes _____ No _X_
Rationale: _____

HYDROLOGY

Is the ground surface inundated?   Yes _____ No _X_   Surface water depth: _____
Is the soil saturated?   Yes _____ No _X_
Depth to free-standing water in pit/soil probe hole: _____
List other field evidence of surface inundation or soil saturation.

Is the wetland hydrology criterion met?   Yes _____ No _X_
Rationale: _____

JURISDICTIONAL DETERMINATION AND RATIONALE

Is the plant community a wetland?   Yes _____ No _X_
Rationale for jurisdictional decision: _____

[1] This data form can be used for the Hydric Soil Assessment Procedure and the Plant Community
   Assessment Procedure.
[2] Classification according to "Soil Taxonomy."

DATA FORM

Field Investigator(s): __Dr. John G. Lyon__ Date: __5/1/92__
Project/Site: _____ State: __OH__ County: __Franklin__
Applicant/Owner: _____ Plant Community #/Name: __3__
Note: If a more detailed site description is necessary, use the back of data form or a field notebook.

- -

Do normal environmental conditions exist at the plant community?
Yes __X__ No _____ (If no, explain on back)
Has the vegetation, soils, and/or hydrology been significantly disturbed?
Yes _____ No __x__ (If yes, explain on back)

- -

VEGETATION

Dominant Plant Species	Indicator Status	Stratum	Dominant Plant Species	Indicator Status	Stratum
1. Ulmus rubra	FACW+	Tree	11.		
2. Acer saccharun	FACU-	Tree	12.		
3. Acer saccharium	FACW	Tree	13.		
4. Rubus allegheniensis	FACU-	Ground	14.		
5.			15.		
6.			16.		
7.			17.		
8.			18.		
9.			19.		
10.			20.		

Percent of dominant species that are OBL, FACW, and/or FAC __55__ %
Is the hydrophytic vegetation criterion met? Yes __x__ No _____
Rationale: _____

SOILS

Series/phase: __Bennington__ Subgroup:[2] _____
Is the soil on the hydric soils list? Yes _____ No __x__ Undetermined _____
Is the soil a Histosol? Yes _____ No __X__ Histic epipedon present? Yes _____ No __X__
Is the soil: Mottled? Yes __X__ No _____ Gleyed? Yes _____ No __x__
Matrix Color: __10 YR 5/4, 5/4__ Mottle Colors: __10YR 6/5__
Other hydric soil indicators: _____
Is the hydric soil criterion met? Yes __x__ No _____
Rationale: _____

HYDROLOGY

Is the ground surface inundated? Yes __x__ No _____ Surface water depth: _____
Is the soil saturated? Yes __x__ No _____
Depth to free-standing water in pit/soil probe hole: _____
List other field evidence of surface inundation or soil saturation.

Is the wetland hydrology criterion met? Yes __x__ No _____
Rationale: _____

JURISDICTIONAL DETERMINATION AND RATIONALE

Is the plant community a wetland? Yes __x__ No _____
Rationale for jurisdictional decision: _____

_____

[1] This data form can be used for the Hydric Soil Assessment Procedure and the Plant Community Assessment Procedure.
[2] Classification according to "Soil Taxonomy."

DATA FORM

Field Investigator(s): _Dr. John G. Lyon_   Date: _5/1/92_
Project/Site:_____ State:_OH_ County: _Franklin_
Applicant/Owner:_____ Plant Community #/Name:_____4_____
Note: If a more detailed site description is necessary, use the back of data form or a field notebook.
- -

Do normal environmental conditions exist at the plant community?
Yes _X_ No _____ (If no, explain on back)
Has the vegetation, soils, and/or hydrology been significantly disturbed?
Yes _____ No _X_ (If yes, explain on back)
- -

VEGETATION

Dominant Plant Species	Indicator Status	Stratum	Dominant Plant Species	Indicator Status	Stratum
1. Typha latifolia	OBL	ground	11.		
2.			12.		
3.			13.		
4.			14.		
5.			15.		
6.			16.		
7.			17.		
8.			18.		
9.			19.		
10.			20.		

Percent of dominant species that are OBL, FACW, and/or FAC _____100_____ %
Is the hydrophytic vegetation criterion met? Yes _x_ No _____
Rationale: _____

SOILS

Series/phase:_____ Pewamo _____ Subgroup:[2] _____
Is the soil on the hydric soils list? Yes _X_ No _____ Undetermined _____
Is the soil a Histosol? Yes _____ No _X_ Histic epipedon present? Yes _____ No _X_
Is the soil: Mottled? Yes _____ No _X_ Gleyed? Yes _____ No _X_
Matrix Color: _10YR 2/2, 2/2_ _____ Mottle Colors: _____
Other hydric soil indicators: _____
Is the hydric soil criterion met? Yes _x_ No _____
Rationale: _____

HYDROLOGY

Is the ground surface inundated? Yes _X_ No _____ Surface water depth: _0.5 "_
Is the soil saturated? Yes _x_ No _____
Depth to free-standing water in pit/soil probe hole: _____
List other field evidence of surface inundation or soil saturation.

Is the wetland hydrology criterion met? Yes _X_ No _____
Rationale: _____

JURISDICTIONAL DETERMINATION AND RATIONALE

Is the plant community a wetland? Yes _x_ No _____
Rationale for jurisdictional decision: _____

[1] This data form can be used for the Hydric Soil Assessment Procedure and the Plant Community Assessment Procedure.
[2] Classification according to "Soil Taxonomy."

Bibliography

Agricultural Research Service, 1971. *Common Weeds of the United States*. U.S. Department of Agriculture, Washington, D.C., and Dover Publ., New York, 463 pp.

Billington, C., 1968. *Shrubs of Michigan*. Cranbrook Institute for Science, Bloomfield Hill, MI, 339 pp.

Braun, E. L., 1964. *Deciduous Forests of Eastern North America*. Hafner, New York, 596 pp.

Braun, E.L., 1989. *The Woody Plants of Ohio*. Ohio State University Press, Columbus, 362 pp.

Britton, N. and A. Brown, 1970. *An Illustrated Flora of the Northern United States and Canada*. Volumes I, II, and III, Dover Publ., New York.

Cobb, B., 1963. *Ferns*. Houghton Mifflin, Boston, 281 pp.

Courtenay, B. and J. Zimmerman, 1972. *Wildflowers and Weeds*. Van Nostrand Reinhold, New York.

Dana, W., 1963. *How to Know the Wildflowers*. Dover Publ., New York, 418 pp.

Fassett, N., 1975. *A Manual of Aquatic Plants*. University of Wisconsin Press, Madison, WI, 339 pp.

Federal Interagency Committee for Wetlands Delineation, 1989. Federal Manual for Identifying and Delineating Jurisdictional Wetlands. U.S. Government Printing Office, Washington, D.C., 76 pp.

Fisher, T., 1988. *The Dicotyledoneae of Ohio, Part III, Asteraceae*. Ohio State University Press, Columbus, 280 pp.

Forey, P., 1990. *Wildflowers*. Gallery Books, New York, 239 pp.

Foth, J., 1990. *Fundamentals of Soil Science*. John Wiley, New York.

Harlow, W., 1946. *Fruit Key and Twig Key*. Dover Publ., New York, 126 pp.

Harlow, W., 1957. *Trees of the Eastern and Central United States and Canada*. Dover Publ., New York, 288 pp.

Hitchcock, A., 1971. *Manual of the Grasses of the United States, Volumes I and II*. U.S. Department of Agriculture, Washington, D.C., and Dover Publ., New York, 1051 pp.

Hotchkiss, N., 1972. *Common Marsh Underwater and Floating-leaved Plants of the United States and Canada*. Dover Publ., New York, 124 pp.

Hutchinson, G., 1975. *A Treatise on Limnology, Vol. III, Limnological Botany.* John Wiley, New York, 660 pp.

Knobel, E., 1977. *Field Guide to the Grasses, Sedges and Rushes of the United States.* Dover Publ., New York, 83 pp.

Knobel, E., 1988. *Identify Trees and Shrubs by Their Leaves, a Guide to Trees and Shrubs Native to the Northeast.* Dover Publ., New York, 47 pp.

Little, E., 1979. *Forest Trees of the U.S. and Canada.* Dover Publ., New York, 70 pp.

Little, E., 1980. *The Audubon Society Field Guide to North American Trees.* Knopf Publ., New York, 714 pp.

Lyon, J. 1987. Maps, aerial photographs and remote sensor data for practical evaluations of hazardous waste sites. *Photogramm. Eng. Remote Sensing* 53:515–519.

Lyon, J., R. Drobney and C. Olson, 1986. Effects of Lake Michigan water levels on wetland soil chemistry and distribution of plants in the Straits of Mackinac. *J. Great Lakes Res.* 12:175–183.

Lyon, J. and R. Drobney, 1984. Lake level effects as measured from aerial photos. *J. Surv. Eng.* 110:103–111.

Mitsch, W. and J. Gosselink, 1986. *Wetlands.* Van Nostrand Reinhold, New York, 539 pp.

Mitsch, W. and S. Jorgensen, 1989. *Ecological Engineering, and Introduction to Ecotechnology.* John Wiley, New York, 472 pp.

Mohlenbrock, R., 1987. *Wildflowers, a Quick Identification Guide to the Wildflowers of North America.* Macmillan, New York, 203 pp.

Mohlenbrock, R. and J. Thieret, 1987. *Trees, a Quick Reference Guide to Trees of North America.* Macmillan, New York, 155 pp.

Peterson, R. and M. McKenny, 1968. *A Field Guide to Wildflowers of Northeastern and North-central North America.* Houghton Mifflin, Boston, 420 pp.

Petrides, G., 1988. *A Field Guide to Eastern Trees, Eastern U.S., and North America.* Houghton Mifflin, Boston, 272 pp.

Reed, P., 1988. National List of Plant Species that Occur in Wetlands: National Summary. U.S. Fish and Wildlife Service, Biological Rep. 88 (24), 244 pp.

Sargent, C., 1969 . *Manual of the Trees of North America.* Volumes I and II, Dover Publ., New York.

Smith, H., 1966. *Michigan Wildflowers*. Cranbrook Institute for Science, Bloomfield Hills, MI, 468 pp.

Spencer, E., 1957. *All about Weeds*. Dover Publ., New York, 333 pp.

Symonds, G., 1958. *The Tree Identification Book*. W. Morrow, New York, 272 pp.

Symonds, G., 1973. *The Shrub Identification Book*. W. Morrow, New York, 379 pp.

Trelease, W., 1967. *Winter Botany, an Identification Guide to Native Trees and Shrubs*. Dover Publ., New York, 83 pp.

U.S. Army Corps of Engineers, 1987. Wetlands Delineation Manual. Tech. Rep. Y-87-1, Department of the Army, Washington, D.C.

U.S. Department of Agriculture, 1962. *Soil Survey Manual*. Soil Conservation Service, Washington, D.C., 502 pp.

U.S. Department of Agriculture, 1971. Common Weeds of the United States. Dover Publ., New York, 463 pp.

U.S. Department of Agriculture, 1991. Hydric soils of the United States. Soil Conservation Service, Misc. Publ. 1491, Washington, D.C., 502 pp.

Voss, E., 1972. *Michigan Flora, Part I: Gymnosperms and Monocots*. Cranbrook Institute of Science, Bloomfield Hills, MI, 488 pp.

Index

151